About the Author

Trevor Norton, whose first book, *Stars Beneath the Sea*, won him the accolade 'Bill Bryson underwater', is Professor of Marine Biology at the University of Liverpool and Director of the Port Erin Marine Laboratory on the Isle of Man. He has authored over 150 scientific publications and books, and is an authority on the history of scientific diving.

Also by Trevor Norton

Stars Beneath the Sea (Century, 1999)

Trevor Norton

Reflections on a
Summer Sea

Illustrations by Win Norton

ARROW

Published by Arrow Books in 2002

7 9 10 8

Copyright © Trevor Norton 2001

The right of Trevor Norton to be identified as the author of this
work has been asserted by him in accordance with the
Copyright, Designs and Patents Act, 1988

First published in the United Kingdom in 2001 by Century

Arrow Books Limited
The Random House Group Limited
20 Vauxhall Bridge Road, London SW1V 2SA

www.rbooks.co.uk

Addresses for companies within
The Random House Group Limited can be found at:
www.randomhouse.co.uk/offices.htm

The Random House Group Limited Reg. No. 954009

A CIP catalogue record for this book
is available from the British Library

ISBN 9780099416166

The Random House Group Limited supports The Forest Stewardship
Council (FSC), the leading international forest certification organisation.
All our titles that are printed on Greenpeace approved FSC certified paper
carry the FSC logo. Our paper procurement policy can be found at:
www.rbooks.co.uk/environment

Typeset by SX Composing DTP, Raleigh, Essex
Printed and bound in Great Britain by
CPI Antony Rowe, Chippenham, Wiltshire

To Jack and John,
who befriended and inspired
several generations of young ecologists

Though memory's tide soon runs to neap,
The Lough remains, dark and deep,
And I have promises to keep,
To tell the tale for those who sleep.

With apologies to Robert Frost

Acknowledgements

My story is, above all, a personal history, but I am indebted to all those too numerous to mention who shared their memories and photographs. None the less, without the words of John Bohane, John Ebling and Jack Kitching this would have been a dull narrative.

I am indebted to Maíréad Nic Craith who ensured that my Irish was Irish.

I am also deeply grateful to John Bishop, Caroline Dawnay, Cathy Kennedy, Colin Little, Rachel Norton, Peggy Pertunnen, and Arthur and Phyllis Pollard, who all commented on parts of the first draft.

My greatest appreciation is, however, reserved for my wife, Win, for more than I can ever express, plus the drawings that adorn these pages; Annabel Hardman, who fell in love with a Lough she had never seen; and Mark Booth who may, if he perseveres, make a writer out of a raconteur.

Credits

I would like to thank all those responsible for giving permission to reproduce extracts from the following copyright material:

T. Barry, *Guerilla Days in Ireland* (1949). Reprinted by permission of Anvil Books.

Eamon de Valéra's radio speech on St Patrick's Day, 1943. Quoted by permission of the Archive department of Radio Telefis Eireann.

T. Lehrer, *An Irish Ballad*. Quoted by kind permission of Tom Lehrer.

A.C. Mathews, 'Death of the Irish', from *Minding Ruth* (1983). Reprinted by kind permission of the author and The Gallery Press, Loughcrew, Oldcastle, Co. Meath, Ireland.

Sean O'Faolain, *An Irish Journey* (Longmans, Green & Co., 1941) Copyright © as in the original edition. Reproduced by permission of the estate of Sean O'Faolain c/o Rogers, Coleridge & White Ltd., 20 Powis Mews, London WII IJN.

E. Sommerville & M. Ross, *Some Experiences of an Irish R.M.* (Longmans, Green & Co., 1906). Reprinted by permission of Curtis Brown.

While every effort has been made to secure permissions, I apologise for any apparent negligence and undertake to make any necessary corrections in future editions.

Preface

Professor J. A. Kitching OBE, FRS

24 Oct. 1994

Dear Jack,

I have begun to write the story of Lough Ine. I want to tell of the stunning scenery and terrible history of the place, the myth and the magic, and to recapture all the fun and excitement we had in those summers when we waded and dived in the lough. Perhaps I can convey the wonder I felt when I first came to Lough Ine in the 1960s, and maybe slip in a bit of marine biology too . . .

This is the story of the menagerie of eccentric and talented ecologists who, as a hobby, established a privately owned field laboratory in south-west Ireland and took part in one of the most unlikely projects in the history of marine biology. All the characters are real and the tales are as true as memory will allow. I was present when many of the incidents occurred, for I spent fourteen wonderful summers at Lough Ine learning to be an ecologist and a second-string eccentric. Others, who were there before or after me, have supplied the events I missed. They demonstrate that although science is a serious, rational business, scientists can be seriously frivolous and irrational people.

It is also a biography of the lough, an inlet of the sea masquerading as a lake, and one of the most renowned ecological sites in the world. I have called it Lough Ine (pronounced as in wine), the name by which it became famous. Now, for no good reason, it has become Lough Hyne. Of course, many a place name in Ireland was enshrined on the map because the English Ordnance Survey man misheard the locals: 'The wood of the Berries' (*Fiodh na gCaor*) became 'Vinegar' because that's how it sounds. But Lough Hyne makes no sense in Gaelic. It is not a rational reversion to the lough's original name, for no one knows what it was. Was it the deep lake, Loch Domhain (pronounced dine), or the lake of the whirlpool, Loch Oighin (ine), or St Ina's lake after the saint venerated at a holy well nearby? To clarify the issue, local signposts direct the bewildered traveller to Lough Aidead, Lough Aidhne, Lough Oighin, Lough Hyne and Lough Ine. So let's settle for Lough Ine, because that is how I've always known it.

My story also records a time, only yesterday yet long ago, before grants and subsidies catapulted rural Ireland into the 21st century and brought prosperity at last to the most hospitable people on earth. It was a time before Ireland became a land of roadworks, with every village renamed Loose Chippings; before the house plans in Fitzsimon's *Bungalow Bliss* stimulated thousands of otherwise decent people to erect flashy haciendas without a trace of an Irish accent. As a local admitted, 'What with balustrades and balconies we've been bungalowed to buggery.'

It was a time when much of the far west, that receptacle for rocks, was a wilderness of inspiring beauty, before the wild moors were scraped black by machines to feed peat-burning power stations, and before the landscape was obscured by thickets of accommodation boards. I overheard someone plead, 'Please, God, give us a sign . . . that doesn't say Bed and Breakfast.'

'Ah sure,' I was told, 'it has all changed now. But even if it hadn't, it would still be different.'

It was only yesterday, when I was young and had no idea of the strange adventure about to befall me.

Arrival

There are no snakes in Ireland, but the roads make up for it. The one from Skibbereen slithers over and around the terrain as it tries, but fails, to find an easy route. Everywhere there are hillocks and outcrops and bogs, but in West Cork it is not the scraped-bare, soilless desolation of some parts of Ireland, where the ominous feeling is that one day the ice will return. Here there is a greener barrenness, but don't be fooled by the honeysuckled hedgerows or the emerald veneer on the fields that promise fertility. The dolphin-backed outcrops are just the few that break the surface; great masses of bedrock lurk below. It is a harsh, unyielding land that wears a man down and into an early grave in soil too shallow to dig.

In 1964 cars had yet to surpass the cow as king of the road, but were common enough to be a menace, for there was no speed limit on the highway and not everyone had got the hang of how to drive. An official driving test was still a year away and there would be no retrospective assessment for drivers already careering down the road. A survey of those involved in accidents had alerted me to the perils of hats on the highway. Beware, it said, of the driver with the determined look and the trilby, worse still the one with the cap and the vacant stare — but above all avoid the nun with her blinkering cowl and the certain knowledge that God will protect.

Yesterday I had seen Liverpool slip away into the gathering darkness as the overnight boat left for Dublin. Ferries weren't floating hotels in those days. This tub was kept afloat by its barnacles. Everything was painted black, layer upon layer over pustules of rust. I had read that 125 years earlier the *Irish Packet* possessed 'a miserable paucity of accommodation and utter indifference to the comfort of the passengers'. I think it was the same boat.

On deck, the young couple across from me were sitting on a bench holding hands — she in a funny hat, he in his father's suit dandruffed with confetti. Feeling as if I were sitting in their bedroom, I left to explore the ship.

The saloon bar was the warmest place on board and full of smoke, the ceiling tanned with nicotine, the windows opaque, the curtains stiff and brown. We would all be kippered by morning, I thought with a smile.

I chatted to some well-washed American students with new-mown crew cuts — a sample of the 38 million Americans who claim Irish ancestry. They came from Boston, convinced they were Irishmen coming home. That's the trouble with Ireland: it steals the heart and confuses the passport. There is an aching loveliness in the country that hauls back not only those who have left, but those whose great grandfathers left. Ireland, they say, is a fatal disease.

A group of English teenagers drank too much and sang songs for which the lyricists seemed only to have written the first verse. Their version of 'The Wild Rover' was as loud and out of tune as usual.

As the evening wore on, the teenagers paired off and went out on deck. Eventually I followed and, after the smoke, the fresh air made me cough. I walked back along the starboard rail and looked over the long aft deck at the luminous wake trailing into the night. As my eyes became accustomed to the gloom, I noticed strange movements from the wooden life rafts on the deck below. On each one a pallid bottom was rising and falling in the moonlight. A star winked at me and I winked back.

I settled down for the night on the draughty deck. What on Earth was I doing here, sleepless in the middle of the Irish Sea? It was Dr Burrows' fault.

As an undergraduate I had correctly estimated the minimum amount of work required to get a good degree. Surprisingly my mentor, Dr Elsie Burrows, decided I was worth retaining and secured a studentship for me to do ecological research for three years under her supervision. I was flattered and accepted, although I thought the project more worthy than exciting. I was to unravel the patterns of growth and reproduction in a commercially exploited kelp plant called *Saccorhiza*, an impressive six-footer that lived below the tides and would therefore provide a reason to go diving. The work didn't sound too arduous, and I reckoned I could fit it in when I wasn't playing football.

But most of all I loved the *idea* of becoming a marine ecologist, although I wasn't entirely sure what they did.

I was troublesome from the outset, and Dr Burrows was becoming impatient with me. 'Norton,' she announced, 'I am sending you . . . (was she about to say "down"?) to Ireland. Professor Kitching, who owns the laboratories at Lough Ine, has agreed for you to spend the summer there beside the densest forest of *Saccorhiza* you will ever see.'

So that was that. I hastily read a few of Kitching's publications. They were excellent pieces of research, but referred back to his earlier studies from the 1930s; that meant he must be at least in his fifties by now, probably over the hill. Even so, I might learn something.

I had more to learn than I imagined.

Early next morning the ferry eased into the narrow, stout-stained river Liffey and was lassoed to Ireland. I had arrived in Dublin with empty pockets of expectation, and was bruised and buckled from a miserable night on deck.

What was I going to do for three hours before my train left for Cork? I walked the streets, past doorways of sleeping men, and ragged children begging on street corners.

I then came across a busker. He juggled, but the balls bounced off in all directions; his card tricks failed and his balancing dog fell over. 'And now for somet'ing special,' he declared, producing a large sword and whacking it against his suitcase to show it was steel. He bent back his head, opened his mouth and began to insert the blade. I couldn't watch. Even if he got it down without mishap, which seemed unlikely, he might forget and bow to the audience. I tossed a few shillings into his cap to help with the hospital bills and took refuge in the Guinness brewery just across from the railway station.

In 1759 Arthur Guinness had come into a small inheritance. I would have gone out and got drunk, but Arthur took a 900-year lease on a brewery and got the whole of Ireland drunk. The brewery paid a quarter of all the excise revenue levied in Ireland. It sold best in Africa, where Guinness was considered an aphrodisiac. Not so in Ireland apparently for, according to a local journalist, an Irishman is the only man in the world who would climb over a dozen naked women to get to a bottle of stout.

A tour of the brewery instructed me in the mysteries of double porter, roasted barley and bottom fermentation. These details induced a mighty thirst to put my new-found knowledge into practice in the tasting room. Unfortunately, my fellow tourists kept asking questions and the guide answered at such length that I had to sprint for my train to Cork before the tour had finished. The brewery produced five million pints a year and I couldn't even get one.

I caught the train and, arriving in Cork, rushed through the town with just enough time to get from the railway to the bus station, from where I was due to get the coach for Skibbereen. In a square beside the depot a tramp clutching an empty whiskey bottle snoozed at the feet of the statue of Father Mathew, founder of the temperance movement.

The bus carried poultry, post and passengers.

'Room for one more?' I asked.

'There is ample accommodation within,' the driver replied.

The bus pulled out and had to weave between the crowds. The arms of a policeman on point duty traced elegant arabesques in the air, but no one paid the slightest attention. 'In Cork,' the driver observed, 'the populace move across the streets with as much regard for traffic as sheep on a mountain pass.' In Ireland even bus drivers were orators.

Before we had cleared the city limits the hoardings convinced me I was now in a foreign land and that all Irish roads led resolutely to the past:

THE CORK CORK COMPANY, CORK

GOODWEAR LTD

PEEK-A-BOO REGISTERED

THE SICK AND INDIGENT ROOMKEEPERS' SOCIETY

THE INVALID REQUISITE AND BABY CARRIAGE COMPANY

My fellow passengers energetically crossed themselves at every church and cemetery we passed. 'It is just our way of keeping fit,' said the man sitting next to me, with a wry smile.

This was the most strict Catholic country in Europe. A bishop had denounced the mention of 'nighties' on television and demanded that dangerous pursuits such as late-night dancing should be banned. Ten thousand books *were* banned for being too explicit – 350 in 1964 alone. In response to sexual queries, a magazine agony aunt replied that the answer to the reader's problem would be given by her confessor and, in any case, outside marriage it was unnecessary to know the details.

London was swinging, Cork was dangling in the darkness.

The bus wound its way through towns with magical names: Bandon, Dunmanway, Clonakilty, known as 'Clonakilty God help us', and Leap, once at the limits of civilisation – 'Beyond the Leap, beyond the law'. I browsed through the *Cork Examiner* in which each town had its own little section: 'Clonakilty Cuttings', 'Bandon Brieflets'. One morsel under 'Bandon Brieflets' concerned a priest who had taken ladies' panties from clothes lines. 'This behaviour is not altogether unusual,' the magistrate admitted, 'but it must not be allowed to get out of hand.'

Skibbereen is the centre of West Cork and not a town to be taken lightly. When it was mooted that the cinema should be allowed to open on Sundays, a councillor declared he was 'not going to see Skibbereen turned into another Paris'.

By 1964 Skibbereen, in its long evolution towards modern Paris, had reached Dodge City. In the narrow streets, dusty dogs yawned among the dried cowpats and carts were still more common than cars. Piebald ponies dozed at hitching posts while donkeys burdened with huge panniers dawdled along the street gazing into the shop windows. Behind the Victorian façades prospectors might be filing their claims, and in each saloon a brawl might be only an insult away. It had the air of a frontier town where any minute something unexpected might happen, only it never did.

I found a bar. It wasn't difficult, for Skibbereen, the home of the first temperance society in Europe, now boasted one pub for every ninety inhabitants, almost four times more generous than the national average. Donelan's bar was a pub untouched by time. It was tiny, with only six stools and a short wooden settle. Twenty customers and it would have been packed. Fortunately there was only one, an old man with a crumpled face and fingers as nicotined as his tooth. He was drinking tea with an air of quiet resignation as if it were his last act on earth. He glanced at me and raised an eyebrow in what appeared to be restrained surprise that a young person had entered. He then turned away in case I became unexpectedly exhuberant.

'Do you have Guinness?' I asked Mrs Donelan, who was standing behind the bar.

She looked at me as if I had gone quietly mad. 'Well I'll have to go search the back room. We don't get much call for Guinness.'

Two pints later, Mrs Donelan pointed me in the direction

of Lough Ine. It was almost five miles, and since I had no money to spare for a taxi I would have to walk.

I found myself trudging down the road past fields misted with summer cobwebs. In spite of the gentle breeze I was sweating and my backpack grew heavier with every stride. I had just dodged my second motoring novice, for this was Ireland, supplier of nuns to the world.

The road seemed never-ending but at last it dropped down. I rounded a bend and suddenly, as if a curtain had been withdrawn, there was the steep wooded shoulder of a hill and the beautiful lough, alive in the sunlight. An island with a ruined castle floated in its centre. The trees came down to the water's edge and on the still surface Cézanne had painted them upside down. I knew at once that this was a magical place with fairies, holy wells and crocks of gold.

Across the water I could see the camp, its tent cones just peaking over a shimmer of leaves. I walked down the western shore where the hill dropped sheer into the lough, and found a gate and a path leading to a small quay. There I changed into my bathing trunks and slipped into the cool waters of Lough Ine for the first time.

It was a ten-minute swim to the camp. A tall, wiry figure was standing on the dock teaching students how to tie sheet bends and rolling hitches. He stared as I rose from the water festooned with seaweed. I knew who it must be.

'Professor Kitching, I presume.'

'King Neptune, without doubt,' he replied. 'Welcome to the Glannafeen laboratory.'

I was tired, wet and weedy – but I had arrived.

Introductions

Professor Kitching looked much fitter than the average fifty-six-year-old. He was tall with long lean legs, and would soon be tanned to the colour of an expensive violin. A swatch of white hair perched on his head and he wore tatty shorts and a sweater with large pieces missing as if it had been savaged by sharks. In spite of his impersonation of a shipwrecked mariner, I knew instinctively that he was the Lord of the Lough.

'Call me Jack,' he said in a voice like a distant foghorn softened by mist.

He was not what I had expected. Professors were not like this. Especially as he wasn't even a run-of-the-mill prof but a Fellow of the Royal Society, the highest accolade that British science can bestow. I felt sure that all term long on campus he wore a jacket and trousers and everyone said, 'Yes, sir,' and, 'Excuse me, Professor,' and that he expected no less. But amazingly he had stepped off the boat in Ireland, put on a pair of faded shorts and exclaimed, 'Call me Jack.'

Dr Burrows had despatched me with the command, 'Just get on with your own work and keep out of Professor Kitching's way. He will stand no nonsense.' Oh dear, I had thought, nonsense was my speciality.

Jack led me up the narrow field to the laboratory, which was little more than a large hut flanked by two big tanks to catch the rainwater. Inside, the only equipment on show was an old microscope that Darwin would have discarded, and an unenclosed centrifuge (a device that used centrifugal force to separate mixtures), which spun like a propeller when you turned the handle – surely designed to propel metal projectiles through the human heart. The shelves were crowded with vials and jars that were filled with mysterious liquids reminiscent of urine.

The adjoining mess room was more impressive. It had a pine floor and ceiling like a Norwegian youth hostel. A lifeless body hung from the open trapdoor in the ceiling.

'My old diving suit,' said Jack. 'A dangling corpse keeps intruders away.

'And this,' he said in the friendly tone of a Victorian surgeon showing his students a particularly gruesome specimen, 'is my colleague, John Ebling.'

Doctor Ebling was forty-six years old and was lounging in a chair with his hair falling over his brow. In contrast to Jack's tanned toughness and 'Let's trudge through the bracken' look, John resembled a cruise passenger who was undecided whether to go up on deck lest the weather became inclement. Whereas Jack's wardrobe would have benefited from a trip to Oxfam, John wore a neat black sweater, sharply creased slacks and new blue deck shoes. He sprang to his enormous feet in an uncoordinated clatter, shook me vigorously by the hand and in a hoarse explosion of a voice said, 'Welcome to Lough Ine. It's marvellous to meet you.' He was a bit overpowering, but made me feel as if he had been waiting for years for this momentous occasion to arrive. I had only been here for five minutes, yet I could see the attraction of being in the company of this pair of eccentric sages.

John was almost as tall as Jack but didn't look it, for while Jack was drawn out like a glass tube over a Bunsen burner, John had a more nourished look. Having almost certainly lacked a chin for much of his life, he was now developing two.

The mess room had a couple of cookers down the left-hand side. Now he was on his feet John went over to supervise the student cook. He sampled the soup, swilling it around his mouth as if testing a fine burgundy. 'A touch more paprika, I think.'

'We need more grit in our rations,' Jack complained.

'You have to realise, Trevor,' John said, 'Jack doesn't dine, he merely refuels for the next task. I'm considering hanging his lunch from a hook outside so that he can snatch it without slowing down as he rushes past.'

'Good idea,' Jack agreed. 'Just like the Royal Mail train.'

'Jack's idea of a feast is peanut butter with jam and sausages with golden syrup. I have seen him pour black treacle over fried eggs.'

'Yummy,' said Jack, smiling with pride and enjoying the notoriety of being a consumer of curiosities.

'I act as catering manager,' John explained. 'Taking care of all the ordering and choosing the menus. Most of our supplies are brought in from Cork city by lorry at the beginning of the trip. Among the few things we buy in Skibbereen are crates and crates of Guinness, and boxes and boxes of tinned cat food for crab bait.

The locals are amazed we can keep student morale so high on such limited fare.

'Let's have a Guinness before dinner,' he added, and brought bottles for Jack and me. 'Drinks and chocolate are considered essential nutrients and are available at all times. Just take what you want and tick it off on the card. Sup now, pay later.'

At dinner we ate from huge Denbyware plates with ice-blue faces and nut-brown bottoms. That night shoulder of lamb with boiled potatoes and fresh vegetables were on the menu, followed by a delicious lemon meringue pie.

Everyone was relaxed and chatty, and there was an easy informality between the staff and students. The fifteen under-graduates were a typical bunch of bright and bouncy eighteen-year-olds. Obviously they would have known each other before coming to Lough Ine, although nowhere near as well as they would by the end of the trip. I was the outsider I suppose, but I didn't feel it.

When the meal was over, the rituals began. Much human ingenuity has been devoted to devising methods for making coffee. Lough Ine's contribution was ceremonial. John, the master of ceremonies, explained. 'First, put ten heaped spoonfuls of ground coffee into the jug and add boiling water. Then place the jug in a hot oven for five minutes, remove, stir twice – never three times – and replace for a further five minutes. Only when the handle is too hot to hold is it ready to serve.'

It also had to be strained to remove the sludge, but it was the best coffee I had ever tasted.

Then came the washing-up ritual. Large enamel bowls were filled with boiling water from the urn in the corner, and I do mean boiling. Jack carried a full bowl to the table, fog-horning, 'Hot water! Hot water!' There were two production lines, each with a washer, a rinser and a nest of wipers standing by in anticipation. The water was too hot to put your hand in, so Jack prised up the crockery with the dish mop then, with a lightning snatch of his left hand, dropped it into the rinsing bowl before his fingers melted. By now the plate was sterilised, and so were the rinsers as it splashed into the bowl and scalded their shorts. Often the rinser would gain revenge by shouting 'Reject!' and tossing the plate back into the washer's bowl. Some of the wipers were recirculating the dishes behind Jack's back to be washed again and again, until he noticed. 'Hoy!' he exclaimed, raising his dish mop in ceremonial threat.

Jack clearly enjoyed this sort of nonesense as well as being teased by his students. It was a place where anyone who was lively and a bit daft – someone like me – could be very happy.

After dinner the students and I chatted and laughed together. They were all first-years, indeed, this was the first year of all. Jack was Professor of Biology at the brand-new University of East Anglia, and this cohort would always be the oldest kids in the school with no footsteps to follow, no standards yet set. I don't think they realised what a singular experience they were having.

I don't know how Jack had selected them. Perhaps he had put them through a welter of psychological tests and a commando survival course, and then just picked the ones he liked best. He had chosen well, for they all seemed to get on together. Had they not, being confined in this isolated camp might have been a trial. As it was, we were here for four weeks and, although I would probably never see any of them again, it was going to be fun while it lasted.

Even the oddball student was fun. He was as thin as a Biro and blotchy to boot, and insisted on reciting the entire plot of his favourite science-fiction film: 'And then the aliens turned him into a throbbing, yellow putrescent mass' – a fate that awaited almost every member of the cast. Another lad began his stories with, 'A killingly funny thing happened to me the other day,' and ended, 'Well at least *I* thought so.'

As darkness fell, the candles were brought out in their Wee Willie Winkie holders – cream enamel saucers with cup-handle grips. I discovered for the first time the soft, flickering magic of candlelight and the irresistible pleasure of playing with warm wax.

Eventually we shuffled off to bed, clutching our candlesticks and making tantalising shadow shows on the tent walls. One of the lads threw the girls into panic by dragging a bracken frond on a fishing line across the canvas.

Jack slept alone down by the quay, well away from the other tents which were pitched towards the top of the field beyond the mess hut. I shared with John, who otherwise would have had a tent to himself. He didn't seem to mind at all and kept up a jolly banter about almost everything including Victorian glass and his favourite Donald McGill postcards: 'Picture a floozy with rouged cheeks, in a short skirt and a feather boa. The caption claims: "She never smokes, she never drinks, and only swears when it slips out!"' He roared like a drain in flood. The marvellous thing about John was that he always seemed to be enjoying himself. Even if he was telling a joke for the tenth time, the punchline still took him by surprise and he exploded with laughter. He was the life and soul of his own party.

At midnight we blew out the candle and fell quiet. But we were the only ones who did, as the students still chattered on. They had lowered their voices, so that the babble was audible but we couldn't make out a word. It was as annoying as a tap dripping in Morse code. Finally, John became exasperated and decided to shut them up. He began to sing a bouncy aria from *Rigoletto* at the top of his voice. Within seven bars everyone was yelling, 'Pipe down! Shut up!' So he did, and there was silence. And in the silence we all went to sleep.

Until John's snoring woke us up again.

Genesis

I awoke at dawn, startled by the silence when John suddenly ceased to snore. I felt that I was on the brink of an adventure and not a moment should be missed, so I dressed and went out to explore. It was not a big expedition, for my new home was a narrow field and it would have taken me only five minutes to walk the entire perimeter. The plot was surrounded by scrubby hawthorns, our barrier against the surrounding sea of bracken. The western margin bordered a small inlet called Anchorage, where the boats were kept, and at the narrow northern hem of the field was the concrete quay with the lough beyond.

The sun rose fat and eager to shine. Tugged by a warm wind from the south, the boats fretted at their moorings. Wavelets splintered sunlight on the surface of the lough, and licked the shore as if they would never tire of the taste.

Jack emerged from his tent.

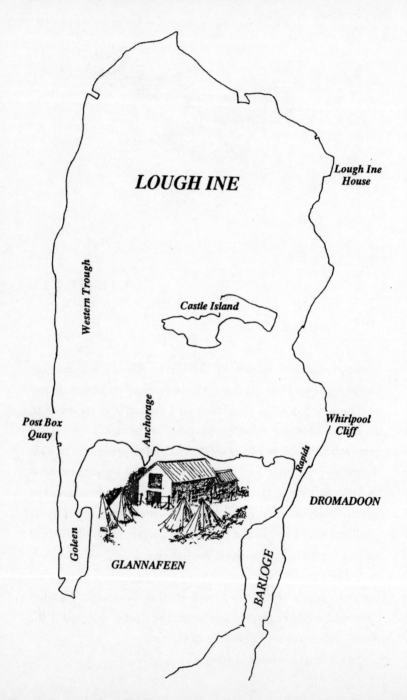

LOUGH INE

Lough Ine House

Western Trough

Castle Island

Post Box Quay

Anchorage

Whirlpool Cliff

Rapids

DROMADOON

Goleen

BARLOGE

GLANNAFEEN

'I'm sorry, I didn't mean to wake you,' I said.

'You didn't,' he replied. 'Are you an early riser too?'

'Not always,' I confessed, as if I had carelessly missed the occasional sunrise. Not always? Not ever. Dawn was a time I associated with getting home from parties, not getting up.

'I have awoken to this view on a thousand mornings,' said Jack, 'yet it never fails to inspire me. Everyone needs a calm place like this in which to anchor.'

'I've always lived in towns. Never been to a place that feels as if it has looked almost exactly the same since time began.'

'Well, since far beyond the reach of memory,' Jack replied, 'but twenty thousand years ago the land was in the grip of perpetual winter. Dublin was over half a mile under the ice. We have retrieved boulders from the lough that were dragged here from as far away as Yorkshire and Scotland.'

I tried to envision a cliff of ice pressing remorselessly southward, planing smooth the rock and grinding it to mud, scouring away the vegetation and soil, and leaving valleys hanging in mid air, their lower reaches lost for ever. It was unimaginable.

'The geologists claim that two small glaciers converged here and plucked a deep depression from the coastal sandstone a kilometre long and three quarters of a kilometre wide. Then, about eight thousand years ago, when the ice became bored with vandalising the land, the melt waters flooded in and formed a freshwater lake.'

I knew from my first-year geology course that the land had sunk beneath the oppressive weight of ice, but when the ice was removed it began to to expand its chest again, straighten its shoulders, and was still doing so. For a moment I thought I could feel the ground rising beneath my feet.

Jack's voice brought me back down to earth. 'Of course a

huge volume of ocean was still locked up in ice caps, and the level of the sea was ninety metres lower than it is today. But as the ice melted, the sea seeped into the basin to make a brackish lake, and then, about four thousand years ago when the sea finally broke through, Lough Ine's tidal pulse began to beat.'

If only I had thought of all this on my journey here, as I had passed through the gaunt landscape where the memory of the ice remained in the form of lakes and soggy marshes. If my train had not been half an hour late but a few thousand years early, I would have seen the lynx and lemmings that flourished in the cold air. I might have caught the tundra giving way to hazel and birch, and eventually the oaks that came to clothe the shores of the lough where wild boars pilfered the acorns.

'Another wonderful, wonderful morning,' exclaimed John Ebling as he strode down to join us on the quay.

'I was telling Trevor how geology has provided us with the ideal site for research,' Jack said. 'A salty lake, sheltered in all weathers, with shallows and deeps where we can do ecological experiments that would be impossible on the open coast.'

'Absolutely,' John agreed. 'In all these twenty-six years our experiments here have never been disturbed by waves, pollution or people.'

'Don't the locals fish in the lough?' I asked.

'Mostly they go out to sea,' John replied, 'if they go out at all. Neolithic men did more damage. They stripped out the oysters, put paid to the brown bear and the great elk, and their axes began to clear the forests for palisades and fuel.'

As we stood together I felt that I was part of a group of old acquaintances, one of whom had to be brought up to date on events since last they'd met.

John clearly loved to expound on Ireland. 'The history of

Ireland is a story of the rape of the land. Cromwell decided that the "barbarous and bloodthirsty" Irish would never be subdued while there were leaves on the trees, so he chopped them down. Then he cut down half the population too. In the seventeenth century oaks were exported to make the wooden walls of England and, in the eighteenth, land was cleared for cattle to satisfy the English demand for beef. Since then, millions of trees have been planted in County Cork to redress the damage; but look around you – most of the hills are still bare.'

'Can't stand here all day,' said Jack suddenly. 'I volunteered to make breakfast.' And he rushed away into the mess hut as if he were late for a train.

'What about a walk down to the Rapids before breakfast?' asked John.

'I'd like that,' I replied eagerly, for I knew from the early publications that the Rapids were just that, a cataract.

'Let me get a stick or something. The path's sure to be overgrown. Nobody has been down it this summer.'

We left by a small gate at the top of the field and struck out to the east through shoulder-high bracken. John went ahead like a jungle explorer, whacking down the plants. He seemed to be wading through a great green surf.

'It used to be all open fields here, but there were dozens of cows in those days. It's desperately poor land of course; half of every farm is useless rock and bog. The walls were built just to clear the soil of stones. Do you know it's against the law to allow thistles to grow in the fields?' John smiled at the very idea.

'It would be a bold thistle that thrives here,' I said.

'When I first came in 1937, the hills were patterned with fields stitched together from scraps of reject material. Nobody in his right mind would have attempted to scrape a living from them, but the locals tried, digging peat and scavenging seaweed from the shore to feed the soil. They grew potatoes on the same plot for a couple of years, then barley, and then the soil was spent, so the farmer tilled elsewhere, leaving the old field to the weeds. Now it's reverting to how it was long ago.'

Around me I could see some long-abandoned ruins and the ghosts of ancient plots. Fields and hedgerows had fallen to meadowsweet and honeysuckle. This was dereliction at its most beautiful, but dereliction none the less.

'Even before the devastation caused by the great famine, it was desolate,' John continued, shifting gear into didactic over-drive. 'A visitor described the utter destitution and abandonment. He claimed that for miles and miles you could see nothing but fields overgrown with reeds and soaked with surface water, vast desolate bogs, cabins of the most wretched kind, scarcely a tree between you and the horizon, scarcely a human being by the way. "I have seen nothing like it," he wrote, "and I can conceive of nothing worse."'

I felt that, had this visitor been John, even in such apparent desolation he would have found some redeeming feature in a landscape he so obviously loved.

'If it was *before* the famine, why were they so poor?' I asked.

'Because Ireland has always had two societies. The ruling classes prospered – Danes, Normans and then the Anglo Irish known as "The Ascendency" – but the ordinary folk had to pay to rent their own soil. It wasn't until 1903 that the Irish acquired the right to become freeholders, and at long last the gentry's picnic on foreign soil was over. But the land continued to be divided between the sons until the pieces were too small to farm. Once houses nestled in every fold of the land. Look at it now.'

After a twenty-minute walk we arrived at the Rapids in the

south-east corner of the lough. It was a channel 150 metres long but only twelve metres wide, connecting the lough to the sea. We were standing on an open, grassy area on the western bank. By contrast, the opposite shore was a steep slope, thick with scrub and with a small, white building, not unlike an old-fashioned railway signal box, that overlooked the Rapids.

'The Dromadoon laboratory,' said John, anticipating my question. 'We built it ourselves.'

'Not much of a rapids,' I said, looking at the feeble flow.

'You're lucky, it's almost slack water and you're in for a treat.'

'What's so special?'

'Ah well, water rushes in and out under the influence of the tide at three metres a second. Amazingly, it drains for twice as long as it flows in because it can't flush out fast enough over the shallow sill lower down the Rapids. A steepening slope develops on the surface of the water and the fall continues as the sea drops to low tide and even as it rises again, until eventually the levels inside and out coincide at low slack water in the lough. And it's about to happen.'

While there was still a current, the surface was ruffled and almost opaque, but as the water stilled it suddenly became transparent to reveal the dense forest of kelp lurking below. For a moment the Rapids held its breath, then, as the flow reversed, the seaweed began to rise out of the water. Huge *Saccorhiza* plants slowly reared into the air like gigantic brown cobras, then flopped over and crashed back into the water to be caught immediately by the quickening current. The water was soon roaring into the lough.

I was exhilarated. 'That was terrific, I've never seen anything like it.'

'Magical,' John agreed. 'Quite magical. Let's go and see what magic Jack has conjured up for breakfast.'

Barloge

'A breakfast fit for a king,' Jack boasted. 'Guinness omelettes.'

Of all the khaki comestibles I had ever tasted, they were the best.

'Can you row?' Jack asked.

'Well . . . yes,' I lied. After all, how difficult could it be to sit backwards and pull two poles back and forth?

'Then you may row me and one of the students to the Rapids.'

'One of the black boats will be free in about ten minutes,' John shouted.

But Jack decided otherwise: 'We can take the brown boat.'

It would probably have been all right if only we hadn't taken the brown boat. It had a mind of its own. It was light with almost no draught, so with even the slightest inequality in the pull of the oars it went into a spin, and any attempt to correct the turn only

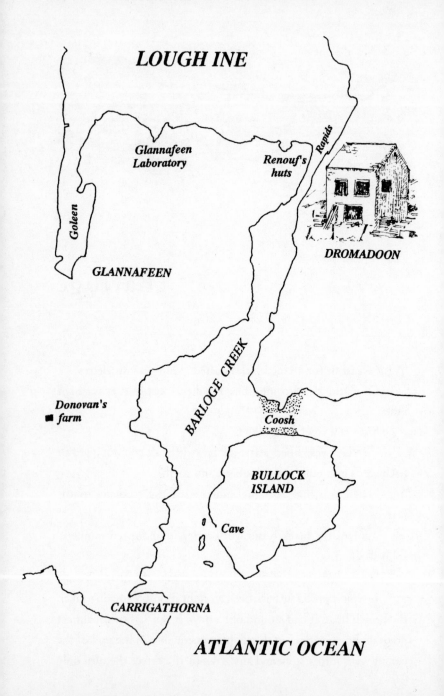

LOUGH INE

Glannafeen
Laboratory

Renouf's
huts

Rapids

Goleen

DROMADOON

GLANNAFEEN

BARLOGE CREEK

Donovan's
farm

Coosh

BULLOCK
ISLAND

Cave

CARRIGATHORNA

ATLANTIC OCEAN

reversed its direction. After we had pirouetted for a while, much to Jack's amusement, I steered an erratic course towards the Rapids.

To make matters worse it was not just *a* brown boat, it was *the* Brown Boat. 'This boat is thirty-five years old,' said Jack. 'It was my twenty-first birthday present. She is a little delicate now, so I don't let just anyone row her. Only expert oarsmen,' he added with a smile.

As we approached the Rapids, I could hear the roar of the water rushing out to the sea. Then I saw the terrifying cataract sucking us towards our death.

'Just reverse the boat and back into Renouf's Bay,' Jack ordered, 'that inlet beside the Rapids.'

Renouf's *Bay*? There was no bay, just a tiny notch in the rock where the water accelerated around the corner into the chaos beyond. With beginner's luck I manoeuvred the boat perfectly into position and backed it gently into the 'bay' in the very teeth of the flow.

'Well done,' said Jack, clearly impressed. I had passed the test.

Jack and Judy, the student, disembarked, and he marched off down the quay that led to the other end of the Rapids.

'Push me off before you go,' I told Judy, who was anxiously looking towards the rapidly receding Jack. I remember that she had wonderful eyes and I was still looking into them when I realised that she had shoved the boat not into the lough, but directly into the current. This was an emergency. As the Rapids yawned before me, with all my strength I heaved on the oars, making them leap from the rowlocks, somersaulting me backwards into the bottom of the boat. I shot down the Rapids sideways with my legs in the air, and by the time I reinverted myself both the boat and I were safely swirling in the turbulent eddies below. A swan sailed gracefully down after me, wondering what all the fuss was about.

'Why, hello,' said Jack sarcastically. 'I didn't expect to see you quite so soon . . .'

A long creek called Barloge leads from the Rapids to the open sea. Barloge means 'top of the weeds' and it is easy to see why, for below the boat were bright green meadows of sea grass over-topped by brown bootlace weed, six metres tall. The shores on either side were fringed with the long, drowned tresses of thong weed.

I rowed out along the creek past a bank of chattering shingle. The iridescent streak of a kingfisher shot into the dark mouth of Bullock Island cave, then out again and away to follow every contour of the shore in level flight. Sprats leaped from the water in arcs to avoid the jaws of a marauding bass, then fell back with the patter of fat raindrops, while mullet sunbathed at the surface tempting the locals to 'fish' for them with a shotgun.

'We can moor here,' said Jack.

In the bow there was a small anchor on top of a rope. Jack pitched the anchor and watched it plummet to the depths. Unfortunately, it was attached to only a metre of rope which, in turn, was unattached to anything on the boat.

'Oh,' said Jack looking mournfully over the side. 'It seems to have gone.'

Fortunately I had my mask and snorkel with me, so I dropped into the water and retrieved it. Judging by Jack's expression, I had managed to redeem myself.

For a while I rowed parallel to the steep shore as Jack scanned the rock in vain for snails. Then I was ordered to hove to, out of the swell, in a gulley spiked with black sea urchins. We disembarked and walked round to the apron of rock facing the sea. Jack went off to gather dogwinkles – snails in quartz-white shells, some striped like humbugs.

The sea had receded and slipped back into itself to gather strength for the turn of the tide. It had exposed a field of barnacles that resembled soiled snow. The tide pools were oases of colour, algal gardens of green tufts on pink carpets, and maroon fans shot with blue iridescence like petrol stains on a damp road.

'Hey, Jude, come and see this.'

Judy bent over the pool I was examining.

'Look at the shape. What does it remind you of?'

'A tide pool.'

'Yes . . . but think big. Look at the long channel running from it and letting the water drain away. Imagine it was the size of a creek.'

'It's just like a model of Lough Ine.'

'Exactly, and it works just like the Lough too. The water's running out through its own little Rapids, and I reckon it'll be half empty before it's refilled by the tide. It's even got bands of microscopic algae around the edge, like a miniature intertidal zone.'

'Maybe we should study the pool instead of the lough,' she said. 'It would be a lot easier.'

'So it would,' said Jack as he rejoined us. 'But even the lough is just a big tidal pool, a system small enough to get to grips with, yet big enough to be representative of the sea.'

'Representative of the sea?' I asked.

'Absolutely. It is wholly marine, but also a confined community with a limited cast of dominant organisms: just one type of urchin and a few species of crabs and starfish. So we can study *all* of them and all their interactions, to reveal how the entire system functions. If we succeed, we will be much closer to understanding the ecology of the shallow seas.'

Although ecologists study nature's complexity, what they seek is simplicity. I hadn't appreciated until now that Lough Ine was indeed a pristine site, uniquely amenable to study, that might serve as a model for the more complex and incomprehensible regions of the sea.

By now the tide had turned and the wind was getting up. 'We had better be heading back to the boat,' said Jack. 'This can be a wild place.'

'An outboard would be useful,' I said.

Jack looked at me as if I had asked for a condom. 'To growl away the tranquillity? Certainly not.'

He had a point, of course. Jack had obviously never thought of rowing boats as inefficient or inconvenient. They were merely 'right' for the place.

'We must keep it just as it is,' he said. 'The way it *should* be.'

I realised that Jack saw himself as the guardian of the lough, its defence against change.

With waves slapping the hull, I took the boat to the mouth of Barloge, where the Atlantic swell piled in to founder on the rocks.

'South from here there is no land until Africa,' said Jack, waving his arm in the vague direction of Morocco. 'When the mood takes it, about every fifty years, the ocean builds great hills of water over twenty-four metres high to challenge the coast. It can be fierce, they say. "A storm so hard it cuts the ears off you."'

Even on a moderate day such as this, the sea leaned upon the land. The impassive promontory of Carrigathorna – 'Rock of

Thunder' – seemed impregnable, but the waves were merely bruised, not deterred, and they were stealing the rock a flake at a time.

It occurred to me that nowhere was immune from change.

Rapids

The next day I persuaded John to return to the Rapids with me, so he would be on hand to haul me to safety if I got into difficulties while diving in the *Saccorhiza* forest I had come to study. I took one of the sturdy black boats this time, and rowed in a straight line despite the stiff cross-wind. Rather than risk trying to moor in Renouf's non-existent bay, we docked in Scyllium (dogfish) bay just short of the corner leading to the Rapids. John leaped ashore to secure the painter, and I clambered after him lugging my diving gear.

Above the shore was a grassy clearing in front of two black, derelict sheds. Their over-long roofs were sagging, the doors were adrift and the side walls slumped to one side. Although the floor was level, the illusion of tilting was so strong that when I strode in briskly I tottered sideways.

'Who do these belong to?' I asked.

'University of Cork. They're Renouf's old labs, abandoned now. One of the locals had warned me that "they is fairly shook and bits are blowing off from time to time."'

I hadn't the time to ask who Renouf was, as low water was less than half an hour away and I reckoned it might be possible to dive in the Rapids for maybe ten minutes on either side of it. That would be sufficient to reconnoitre the underwater forest and perhaps collect a few plants.

As soon as I was kitted up I plunged into the slackening flow. It was fiercer than I imagined, but I clung on to rocks on the bottom and tried to get out of the current. The great canopy of kelp plants above me deflected some of the flow and even though I was only a couple of metres below the surface I was in the dim and disturbing world of a wind-shaken forest.

The current died, and every particle suspended in the water halted for a second as if uncertain where to go. They then reversed and returned from whence they came. All around me the big kelps that had been horizontal in the current rose in slow motion until they were upright and then, as if hit by a sudden gust, they fell. Within moments their great flat stalks were flexed downstream and their blades were trembling in the flow.

As the current grew by the minute I moved carefully by changing handholds on the rock as if I were climbing a horizontal cliff. My knuckles were white from trying to hang on for as long as I could. Underwater it must have been one of the most exhilarating sights in the world: over my shoulder I could see the great plants writhing and flapping like storm-torn flags. My dive mask began to vibrate with increasing violence until at last the accelerating current whipped it off and spat me out into the lough.

On the quay John had enjoyed the drama of slack water

from above. Inspired by the sight, he began to sing at the top of his hoarse baritone. I played accompanying trumpet on my snorkel. After a tumultuous climax that Puccini could never have imagined in his worst nightmares, we bowed to an imaginary audience and graciously extended an arm to acknowledge each other's contribution to the gala performance. We were loudly applauded by some tourists who had arrived unnoticed on the opposite bank and witnessed the entire, bizarre recital.

Back at camp I was full of the spectacle of the reversing rapids and told Jack all about being underwater in the flow.

'Yes, I know,' he said, 'I used to dive a little myself. The

publications may still be on the shelf over there. They are mostly of antiquarian interest now.'

I rummaged through the pile of scientific papers and, sure enough, there were a couple about Jack's early underwater exploits. To my astonishment, the first one was published in 1934. That made Jack the first marine biologist ever to dive in British waters.

As I learned, his equipment had been about as primitive as you could get. The diving bell is just that — a bell-shaped container that could be lowered into the water. Providing it is watertight, the air gets trapped within and, although the water rises up inside, it only does so until the air is squashed to the same pressure as the water at that depth. Jack's helmet was a diving bell made from a small milk churn, with a glass window inserted at the front and a garden hose attached to the side to replenish the air. The helmet simply rested on his shoulders so that the surplus air bubbled out continuously from under the rim and, if he leaned forward, the water rushed in as an invitation to drown.

In 1931 Jack began his underwater exploits by exploring a submerged gully in Devon. His 'diving suit' was a rugby shirt, long shorts and plimsolls. It was difficult to withstand the cold for more than twenty minutes and frequent diving led, he confessed, to a 'greatly increased appetite for sugar and treacle'.

'John, did you know that Jack was a pioneer of diving?'

'Of course. He used to go down in the Rapids in the early days. Started here about 1948 and wore a bloody great bucket on his head. The weights that kept him down were so heavy they could only be attached to the helmet when he was in the water. Getting them on him damn near ruptured me. He also had an intercom, with the microphone covered by a toy balloon to keep it dry. He claimed he was going to dictate notes to a scribe on the surface, but the only message I can remember him transmitting was, "More air! More air!"

'On the surface one of us held the safety line, while another sweated over the air pump and prayed for the moment he would swap jobs. Jack didn't come up until we were all so exhausted we could pump no more. Sometimes we used to make him give up early by easing back on the air.

'He harvested seaweeds from the Rapids with an enormous sickle and hauled up boulders to examine their fauna. He also designed an ingenious miniature diving bell for the boulders so that, one at a time, they could be enclosed underwater, then lifted, after all the water had been gently pumped out, leaving the silt deposits undisturbed. Almost every day for a month he hauled up rocks.'

John suddenly broke off and rummaged through a drawer. 'If I'm not mistaken . . . Jock Sloane, one of the team in the early days, kept a doggerel diary, and . . . Yes! Here's a photocopy.'

He assumed the voice of the poet laureate on a state occasion:

Jack, in his frog suit, wades about
And tries to lift the blighters out.
Sometimes they come, sometimes they stay,
For him to try another day.

The boulders were taken back to the laboratory for analysis of their silt load, or, put more technically:

To wash it off and suck,
To estimate the total guck.

The diving was undoubtedly tiring, but pumping air down to him was a sentence with hard labour. We all cheered when the machine broke down.

No diving could take place. With luck,
The gadget's permanently stuck.

As more and more boulders were collected the whole team got weary and the verse became terser:

8 July   To get some further information,
          We did another boulder station.
9 July   In the morning, once again,
          We bouldered by the quay and then . . .
13 July   Today we really thought we ought
          To gather boulders at site nought
18 July   Jack's diving suit has sprung a leak,
          He found it out in Eddy Creek,
          When he got very cold and wet.
          'Tis sad he had to stop, and yet . . .

19 July   The suit's repaired and we must do
           A good day's work – and Tuesday's too!
21 July   The daily round, the common task.
           What did we do? Well need you ask?

'Jack soon replaced his bucket helmet with a full face mask. It was still supplied by a pump on the surface, but now he couldn't bail out if it flooded. I remember when two students had to haul his head above the surface while we pumped like mad to keep the water at bay. Jack swore through the intercom and made intermittent gurgling noises as the water level rose above his mouth. Another time, he hauled off his face mask too vigorously and his dentures sank to the bottom with a smile – and we smiled too.'

At that moment Jack trundled in and John bawled, 'Hello, Jack! How has the sampling gone?'

'Very well,' Jack replied and smiled broadly.

But all I could think of was his disembodied dentures on the sea bottom, snapping at passing prawns.

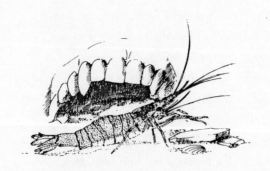

Renouf

I decided that, if there was quiet moment – when Jack wasn't cracking the whip – I would get John to tell me about Renouf. The next afternoon I had finished measuring some plants I had collected and the students were writing up their results in the lab. Jack decided he would row out to test the new temperature probe which was on a long cable that enabled it to reach down into the deepest part of the lough. I offered to go with him, but he politely declined. 'No spare room in the boat with all this cable.'

I therefore bought John a Guinness and asked, 'Who was this Renouf of the leaning labs and the smallest bay in the world?'

'Ah,' he said, 'it's a long story . . .'

All John's stories were long, but this one told how the lough came to be studied.

It had all begun in Easter 1916. With civil war raging in Dublin, the fisheries vessel *Helga* slipped up the River Liffey through the early-morning mist and shelled the rebel head-quarters at Liberty Hall. The first salvo missed and hit a steel bridge with a clang that woke the entire city. The next day one of its guns was mounted on a lorry and sent to pound de Valéra's stronghold in Boland's bakery. Two shells overshot the mark and landed in the river within feet of the *Helga*, drenching her crew. Thinking they were under attack, they fired back, and she came perilously close to becoming the first ship to sink itself with fire from its own guns.

That summer the *Helga* convalesced by cruising the waters off West Cork. Stormbound, she moored in Barloge creek and Richard Southern, a Fisheries officer, devoted five days to collecting specimens from Lough Ine and cramming his note-books with observations. The Royal Irish Academy were impressed by his report and decided that as soon as the Great War ended the area should be investigated in detail. But two long, weary years later it seemed less urgent and nothing was done.

In 1922 Louis Renouf, aged only thirty-five, was appointed to the Chair of Biology at University College, Cork. He was from Channel Island stock and was the grandson of the archaeologist Sir Peter le Page Renouf, an authority on the gods of ancient Egypt, and the first to credit Champollion with having translated the hieroglyphs. Le Page Renouf had converted to Catholicism and served for a time as Cardinal Newman's amanuensis. The cardinal had cradled the infant Louis in his arms.

Louis won a scholarship to King Edward VI Grammar School in Birmingham. With his enthusiasm for living things, he was invited to work in the pathology department at the local university, and another scholarship took him to Trinity College, Cambridge. On graduating, he wrote around seeking a position, and received a telegram from the professor of zoology at

Glasgow: 'Can you begin demonstrating on Monday morning?' So Renouf travelled up on the overnight train, arriving at five in the morning to begin work at eight. For several years he lectured, and was curator of the Hunterian Museum before leaving for Ireland.

The Irish naturalist, Lloyd Praeger, remembering Richard Southern's report, told Renouf of the wonders of Lough Ine. Renouf visited the lough for the first time on a stormy February day in 1923, and remained beneath its spell for the rest of his life.

'What did he look like?' I asked John.

'Oh, he was a sinewy man, tough as leather, with unruly hair and pebble glasses that gave him a half-startled, half-grumpy expression. In his long shorts, a shrunken Fair Isle sweater and a knapsack slung over his shoulder, he had a wild and woolly look.'

I imagined Renouf as a deserter living rough in a John Buchan novel, scavenging crabs from the pools for food, but getting distracted by the complexity of their mandibles and maxillae. He was a man with a secret life.

He bore a passing resemblance to Eamon de Valéra, the Irish President, and at first the locals wondered what pressing matters of state caused 'Dev' to be digging in the mud. When in professorial mode back at the university, Renouf wore a dark cloak and a large floppy hat like a count in a Victorian melodrama.

Renouf was keen to establish a base at the lough, and a bequest to the university supplied the funds. In 1925 he set up a laboratory in one of the old coastguard cottages four miles away in Baltimore, and took timber to the lough to build a hut. Michael Donovan, the local farmer, was shy and suspicious and his wife, who ruled the roost, was antagonistic. She put up a sign that said: GET OUT! and told a neighbour, 'He's a nuisance and a hindrance and I want him off my land.' Renouf soon talked her round by renting from her the dilapidated house beside the farm. He put a large packing case beside the Rapids as a makeshift laboratory until he erected an army hut there, and a few years later two big huts were built just around the corner, the ones I had seen yesterday.

'They were always known as the Cork University Biological Station,' John continued, 'and were constructed by a carpenter from Skibbereen who was also a fine flautist, much in demand to play at wakes. He got so drunk at these that he was incapable of work for a couple of days. If there were two funerals close together the entire week was written off. Renouf used to pray at Mass that no one in Skibbereen would die in the coming week.

'He always kept the most dangerous chemicals on the highest shelf out of reach, but neglected to provide a stepladder.

One day he retrieved a large winchester of sulphuric acid by standing on a rickety cane-bottom chair and the seat gave way. He plummeted downwards and his knees jammed in the narrow frame. He was upright but trapped. His hands still held the winchester tilted above his head. The bung fell out and acid glugged on to his scalp. With smouldering hair he calmly placed the bottle on the floor, wrenched the chair apart and dived through the door and into the lough, emitting a quiet hissing noise. After a brief bath he rose from the waves saying, "Well, that was a little warm around the hat band." Later some of his hair fell out, leaving odd tufts like neglected stooks in a harvested field."

'I don't suppose the labs were very well equipped,' I said.

'Didn't need to be in those days. The sort of marine biology that was fashionable was mostly collecting, sorting and identifying the catch. All that was required were benches and stools – which, if I'm not mistaken, were made from butter boxes – plus microscopes of course, hundreds of bowls and jars, and gallons and gallons of preservative. When you opened the door there was an overpowering smell of formalin and salt-rusted metal. I seem to recall that there was a communal bathing costume, a woollen effort with short sleeves and legs. On skinny Renouf it had broad maroon hoops, but on a rotund lady botanist they were stretched into narrow pink lines.

'There was a well nearby, just below the Rapids. As it was plastered with cow-pats Renouf decided to have the water tested. "Is it safe to drink?" he asked the bacteriologist. "It contains every germ known in Ireland," was the reply. "Both the good and the bad?" "Every one." "Then they will neutralise each other," he concluded, "so that will be fine." And it was.

'Renouf did everything to make guests welcome. In 1929

Kenneth Rees, an enthusiastic young botanist from Swansea, lodged with Renouf and his wife Nora at the lough. There were also four children, two girls and two boys, in a very close and happy family. Nora cooked the meals and it was taken for granted that they would be served only at high tide when the shores were inaccessible for study.

'Three French boats sheltered in the lough during a storm, and in exchange for an ounce of shag and some barm brack – the local currant loaf made with tea – the crew gave Louis enough lobsters and crabs to feed the family for a week. He loved to "hypnotise" a lobster by standing it on its "nose" and stroking its back until it went into a trance and stayed in a handstand.

'The children collected all sorts of insects and amphibia. They kept them in jars in Kenneth's bedroom, so I doubt he slept a wink as they fluttered and scuttled in the dark. Renouf rowed Kenneth around the lough for hours on end every day looking for seaweeds. They used to peer at them through an "underwater telescope" – a conical coal scuttle with a glass window at the wide end. They found lots of other treasures too: a floating coconut from the Caribbean festooned with stalked barnacles, and a packet of love letters lost overboard from an American ship.'

'What was Renouf researching?' I asked.

'I think he was hoping to name everything that lived in the lough. In his first five years here he identified almost fifteen hundred species: sunstars and snakelocks, devil crabs and dead man's fingers.'

That was what most zoologists did in those days. Ecology was still a recent invention, and the vogue was to classify living communities according to the dominant species present and give them names as if they were distinct entities, just as Linnaeus had named species in Latin 200 years earlier. Renouf wrote about marine and freshwater 'associations' and Rees catalogued the inhabitants of the lough's 'saltmarsh subformation', its '*Laminaria* association' and '*Chondrus* society', treating the communities as if they were distinct and static entities rather than dynamic systems whose borders were constantly being adjusted by the environment. Such schemes codified what was there, but didn't shed any light on how the communities worked, or *why* they were there in the first place. It would be Jack and John who would transfer the emphasis from merely describing the patterns of nature to studying the processes by which it was structured.

'You have to realise that it was a time of transition in biological research,' John explained. 'Most topics were becoming experimental, and experiments were, of course, carried out in the laboratory where the conditions could be strictly controlled. But that was too artificial a place to investigate what goes on in the wild. Since agriculturists did field trials of crops, Renouf thought it might be possible to carry out experiments in the calm shallows of the lough, something never attempted before. He planned field experiments, but never got around to doing them. Perhaps he was uncertain whether proper experiments were possible in the sea, or maybe they were just too much trouble.'

'Perhaps he had more important things to trouble him,' I suggested. 'Wasn't Cork a stronghold of the IRA in those days?'

'Yes. Ireland – a stormy sea without calms. Still, Louis seemed as little dismayed by political turbulence as by weather and tide. But you're right, even here the Troubles were never far away. One dark night when the lough was mirror calm and so quiet you could hear a sea anemone sigh, Kenneth saw some French crabbers unloading rifles into a dinghy then vanish quietly into the blackness towards the north of the Lough. The IRA spasmodically attacked local symbols of British rule, such as the new coastguard station, but it was all very civilised. They notified the resident Irish coastguards, arranged a convenient time when they would be out, then burned the building to the ground. As there was nothing else worth burning locally, it all went quiet again.

'At the end of his stay, Kenneth was due to catch a train from Skibbereen on the Monday evening so that he could make passage on the ferry, the *Innisfallen* from Cork the next day. On the Saturday, Renouf's younger son cycled into Skibbereen and ordered a taxi, but it never came, so another was booked for the following Wednesday to catch the next ferry. Kenneth asked the driver why he hadn't come on the Monday. "Ach," he replied, "I was not knowing whether it was last Monday or next Monday you meant, so I erred on the generous side."

'Whenever he was in England, Renouf lectured on the virtues of Lough Ine, and from 1929 to the outbreak of war dozens of parties of researchers and students from twenty or so universities came to the Cork Biological Station. The visitors' book reads like a Who's Who of British marine biology. Julian Huxley and his wife spent the summer of 1933 at the lough decapitating worms and following their regeneration. They came in a Hotchkiss sports tourer with ten horns, which they had christened Alexander the Great. Renouf had booked accommodation for them in Baltimore at an establishment that boasted the only flush lavatory in the district. Maybe so, but it was at the other end of the platform, and the porter kept the key. They were lodged in the flea-infested station master's rooms, and were not amused.

'One of Renouf's daughters told me she remembered nothing of the great man, but never forgot his pretty Swiss wife, Juliette, who wore beach pyjamas with acres of green chiffon blowing in the breeze. "Disgraceful," her mother said. "Quite unsuitable. What will the locals think?"

'Everyone who inscribed the visitors' book was impressed.' John adopted the tone of the rotund lady botanist: '"So many rare organisms aggregated into a single locality. The collecting is truly remarkable, the fauna is certainly the richest I have seen." And it wasn't just marine organisms they studied. There were also enthusiastic reports of the mosses and ferns, the insects and spiders, and the "abundance and persistence of blood-sucking parasites, both wild and domesticated". Others examined the local freshwater lakes, or the pigments in lichens used for dyeing, or they took photographs of cloud formations with a special whole-sky camera.

'I think Renouf's son played stand-off for Munster, and arranged a game between the visiting scientists and a Skibbereen fifteen. Imagine the biologists' horror when they saw a huge billboard:

Skibbereen Rugby Football Club
SKIBBEREEN
versus
COMBINED BRITISH UNIVERSITIES
Kick-off 3pm
Admission 1s 6d

'Skibbereen won by three points, but years afterwards the biologists were still boasting: "Well yes, I played a little rugger in my youth." Then modestly, "My last game was an international for the combined British Universities."

'The Renoufs spent several months each summer at the lough pottering about in boats, collecting and pickling thousands of specimens for further study which rarely took place. Louis subsidised the trips out of the proceeds from his textbooks; they pretty well doubled his salary.'

'Of course!' I exclaimed. I knew I'd heard of him. 'Stork and Renouf's *Fundamentals of Biology* and their *Plant and Animal Ecology.*'

'That's right. Both bestsellers in their day. Those were the happiest times of Louis' life, although he hated to be out of touch with the outside world. As the postman came only intermittently, the girls had to ride a pony into town every day to send and pick up mail. Telegrams were so frequent that the locals assumed his relatives were dropping like flies. "Dear, oh dear," they said sympathetically. "You're after getting a telegram. Is that another one gone?"

'The locals never got the hang of him. They told me that all workmen tried his patience. He described them as "too lazy to spit", and took it for granted that they connived to cheat him. They certainly saw him coming when he was looking for a boat. His first tub was so sluggish and leaky he acquired a long, thin, unstable vessel like a dug-out canoe. It was christened the *Coffin* and did its best to live up to its name. He then acquired a succession of useless craft. In all of them bailing took precedence over rowing. It was a foolish passenger who didn't keep an eye on the shore just in case they needed to swim for it should disaster strike. On one occasion, with three black priests aboard, his boat slowly sank until it settled on the bottom in the shallows. I have never seen a stranger sight than sunken priests sitting with only their heads and shoulders above the water, wide-eyed in wonder at the Lord's mysterious ways.

'Renouf always seemed surprised when his boat sank. He shouldn't have been, for the bung had been lost long ago and replaced by a candle wrapped in a rag. At intervals it would come out with a resigned pop. Water fountained in like an exuberant bidet and, unless the candle was replaced promptly, the boat went down. The vessel had warped thwarts and a sense of humour to match. It preferred to founder in the Rapids or at least in deep water.'

I resolved never to deride Jack's brown boat again.

'Things never quite went as Louis hoped.' John continued. 'His ambitious plans for the laboratories didn't materialise. He had immense charm, but no sense of proportion, and even Ireland wasn't religious enough for him. One year he failed over forty per cent of his students. When they protested, he merely handed out medals of Saint Philomena and told them to pray to her. He even named his house after her and every year he bought a ticket for the Irish Sweepstake and slipped it under her statue.'

'Did she approve?'

'I doubt it, he never won. When the Vatican removed his beloved Philomena from the list of saints he merely said, "I have known her all my life and nothing has changed."

'He could be wonderfully philosophical but he was also impatient and sometimes fell down the stairs because he couldn't be bothered to walk down. He possessed the focused oblivion of an enthusiast. Once engrossed in the denizens beneath a rock, all other worlds ceased to exist. Often he would eventually look up to find he was marooned by the tide or his boat had drifted away and local fishermen had to rescue him.'

'You don't get characters like that nowadays,' I concluded.

At that moment Jack stumbled in through the door, completely swathed in the flex from the temperature probe. 'The cable has got in a bit of tangle,' he admitted. 'I had better buy a reel for winding it in. By the way, there's a terrible stink from the east shore. A big shark has washed in and is rotting on the beach.

What do you think we should do about it?'

'There's only one thing we can do,' said John. 'Let's have it for dinner.' And we both burst out laughing.

That evening Jack was, as usual, late for dinner and, when everyone else was seated, he charged through the mess room shouting 'Coming!' At such times he used his 'I'm hurrying' shuffle with its funny little chopping steps. All heads turned to follow him as he jogged to the wash room, grabbed a bowl, shuffled past the window outside to get water, then back to the wash room. By the time he emerged everyone had stiff necks from swivelling back and forth to follow his antics. It was worse than spending a week at Wimbledon.

'Sorry I'm late,' he said. 'Busy.' As if to imply that those who were on time must have been slacking.

Which, of course, we had.

Water

The next morning Jack made a momentous announcement: 'I don't think it will rain today.'

West Cork is a place where rain comes to retire. The wind travels a thousand miles to reach these shores and, starved of trees to bend, it scours the hills. Thirty miles inland from the coast, sea plantains still flourish in the remnants of wind-blown spray. Weather fronts rush in from the Atlantic and condense as the finest drizzle, so that seventy per cent of the days are what the locals call 'soft'. Almost two metres of rain fall on the coast each year, and twice as much on the hills. It usually lasts for only a few hours, unless there is a storm, when the rain is magnificent, even frightening. But falling torrents is Sunday special rain; the usual stuff is mist with a mission, soft and soaking. I had never heard the old joke, 'There's nothing wrong with our climate, except for the weather,' delivered with such heartfelt sincerity.

Long after the rain has ceased, the spongy land retains the moisture and streams the colour of tea run from the marshes. There are no parched hillsides here, it is amphibious country.

This is not ideal for camping. 'One year,' John told me, 'the rain was so heavy that a tent collapsed beneath the onslaught. Believe me, there is nothing more depressing than the sound of rain beating against the canvas and dripping from the eaves, or the dank smell of decaying vegetation. Every day your clothes grow damper and you have to sit itchy-bottomed all morning on a wooden stool trying to look down a microscope. Some summers it drizzled almost every day and when the shy sun eventually appeared the tents steamed like lazy volcanoes. Everyone would haul out their bedding in an attempt to chase the damp, revealing a surprising array of love-squashed teddy bears, balding fur gloves and blanket scraps – crummy comforters for insecure eighteen-year-olds. In drier summers the showers often waited until we were ready to break camp, and the tents, skirts raised, were airing in the breeze before being folded away with the winter mould. "If it rains," Jack once told us, "you can shelter under the tents, but be sure to take them down first."'

The human body is a salty sea in a bag of skin. Guinness alone cannot replenish this internal ocean, however hard we try. It requires a supply of fresh water.

At Glannafeen we relied on collecting rainwater from the roof into the two large tanks that flanked the laboratory. Even all the winter wetness safely stored was insufficient to sustain us and we relied on the tanks being topped up by summer showers.

But in my first summer at the lough it was very dry, the nearest Ireland comes to drought. Day after day the sun shone and clouds, swollen with moisture, great cathedrals of condensation,

sailed by to burst somewhere beyond, where they could incon-
venience more than just a few campers. The water level in the
tanks fell lower and lower.

The average person requires over two litres of water per day,
and since only dogs drink the stuff neat, we must consume most
of it by other means. At Lough Ine it went mostly for cooking and
making coffee; every pan of potatoes took about seven litres. The
rest was wasted.

Each water tank had a tiny faucet. Jack's conviction was that
large taps wasted water, therefore small ones encouraged frugality.
Unfortunately, he was wrong. With large taps people turn them
on, fill the pan and turn them off, but little taps run slowly and
you lose patience, especially when other tasks are calling. Perhaps
there is just time to do that other little job whilst this miserly tap
is dribbling into the pan. So off you go and, of course, forget
about it. By the time you remember, several panfuls of water have
run to waste. At the end of that arid summer, with the field
parched golden, the grass beside the tank was still long and lush
and emerald green.

When Jack installed a washbasin in the store, the faucet was
not on the sink — for that would encourage waste — it was just
above the floor. The idea was that you filled a bowl from the tap,
then poured it into the basin. Unfortunately, it was the slowest
tap of all and Niagaras of wasted water flowed from forgotten
bowls or cascaded over the side of the too-small sink when you
tipped in the brimming bowlful.

One tank was all but empty. I offered to climb inside and
scrub it down to clean off the threatening crusts of cholera that
clung to the sides.

'No,' said Jack firmly. 'Don't you know we are short of
water? That tank is almost empty. Wait until it is full again, and

then you can do it.' I had no answer to such extraordinary logic.

The well near to the graveyard had long since vanished beneath the brambles, and the lough is as salty as a sailor's tale. A small stream does enter at the north-west corner, but that was too far away to be of use to our droughted camp. Fortunately, water trickles down in the corner of Whirlpool Cliff nearby, and with a chevron of sticks we channelled it into a milk churn. It took almost an hour to fill, and every day for a week a relay of boats had to replenish our stocks.

Other solutions were tried. Sea water was fine for cooking vegetables, but salty coffee was not a success, and socks boiled in sea water were never the same again.

Ablutions consumed little, for we were restricted to a tiny basinful of water for washing and shaving. We normally washed only the bits that showed, yet everyone looked cleaner and smelled sweeter than the average city dweller that jostles past in the street.

The water was cold, even for shaving. 'Frigid water makes your bristles suitably brittle,' Jack asserted, although he never wet-shaved. He used a Thoren's Riviera Clockwork Shaver; even in the old days, I'd been told, he'd had a rust-prone hand-cranked razor that sounded like a mechanical lawnmower croaking for oil.

Jack and I stood side-by-side in the washroom one evening brushing our teeth – mine were still in my mouth. Jack was a great advocate of dentures. 'Troublesome fixtures,' he called home-grown teeth. Each night Jack left his false teeth to grimace at us from a tumbler on the shelf, then shuffled off to bed with the same gummy command: 'Goodnight. Last one to leave turn off the gas and make sure the candles are out.' Someone had put a label on Jack's tooth mug saying: 'Smile and the world smiles with you.' That night his false teeth were replaced with the jaw of a sheep, but next morning, to everyone's disappointment, he appeared wearing his reserve set. The original teeth were returned surreptitiously, and nothing was ever said.

To curb the lads' bad language, the female students laced their toothbrushes with Tide detergent, which wasn't entirely successful in the short term. After much spluttering and spitting the boys emerged foaming at the mouth: 'Which one of you buggers did this?'

Evening tooth-cleaning took place after dark so we carried candles into the washroom and balanced them on a shelf where the mischievous flames stretched as tall as they could until the shelf above was charcoaled.

'Jack used to live in constant dread of the hut catching fire and melting his dentures,' John told me jokingly. 'Ironically, it was Jack who nearly did it. When taking all-night readings of temperature recorders in the loft, he found there were no batteries for the torches. Although candles were strictly forbidden upstairs, there was no alternative. Unfortunately, Jack fell asleep, and woke to find that a row of six candles perched on the roof cross beam had all guttered and the wood was smouldering.'

Beginnings

After dinner and washing up we sat around the table with the students drawing graphs from the day's results. I tried to get Jack to tell me about how he first came to Lough Ine, but he simply said, 'Busy. I haven't the time just now.'

Fortunately, John was rarely too busy to chat. 'I first came in . . . 1937,' he began. 'A year before Jack.'

'How old were you?' I asked.

'Just nineteen. I was a student at Bristol when the prof suggested the students' Zoological Society should go to Lough Ine for their summer field trip. The idea was to do a bit of exploring, collect specimens of interesting species, but most of all to have a good time.

'Getting here wasn't easy in those days. It was standing room only on the Irish Mail from Cardiff so we slipped the porter a ten-bob note to imply that the last coach was to be uncoupled.

When he cried "All change!" the passengers moved to the front of the train and we had the back carriage to ourselves. He even gave us "Reserved" labels to stick on the windows.

'The microscopes were packed in straw, which Irish Customs ritually burned on the dock in case it harboured foot and mouth disease. They also took the serial numbers of every lens for every microscope and threatened to check that each was in its correct box on the return journey.

'The boat was a decrepit tramp steamer more accustomed to cattle than passengers. Then there was the bus from Cork. The driver went like a maniac over the rough roads, consigning all the oncoming traffic to the ditch.

'We all stayed on the west shore in the new bungalow. The bedrooms and even the beds were teeming with wildlife – until a pan of milk was upset on the cooking range and the smoke fumigated the entire house. It was so cold we kept out a bowl of sugar so we could all dip in to stoke our boilers, and I remember a wind-up gramophone with only eight records, two of which were "Donkey Serenade". Oh, and "She was Poor but She was Honest". I think I can still remember the words—'

'What about the place?' I said hurriedly.

'Well, Ellie O'Driscoll owned the bungalow and had a tiny shop at the loughside. She came out each day in a donkey and trap to bring in perishables. In the window were advertisements for five brands of tobacco, but she only stocked Murray's Cut Plug. It was a block as hard as a brick and a student claimed that it tasted like peat, burned like anthracite and damned near set fire to the bowl of his pipe.

'We took turns at cooking. There were brown scones and custard,' he recalled with affection, 'but there were also homogenised horrors of coagulated protein for breakfast. A huge bowl full of junket left to cool on a window ledge was luckily stolen by a mongrel before we ate it. And there was something sinister about the sausages . . .'

'Aside from the menus . . .'

'Oh, soft days and soaking days and rowing in the moonlight.'

'And you worked with Renouf?'

'No, we just rented one of his labs and borrowed a boat. We rigged it up with a great forked tree branch for a mast and a blanket for a sail. I called it the *Junk* – junk by name and junk by nature. The sail carried it briskly in circles or in the wrong direction, so that the distance to be rowed was almost doubled. Apart from that it was fine. Another one of his boats was flat-bottomed and triangular, aspiring to be a grand piano, so we christened it the *Bechstein*. Uhum.'

'Uhum' was John's way of saying, 'Well what do you think about that?'

I was thinking what an unimaginably ramshackle and home-made world it must have been, but I wouldn't fully appreciate the condition of pre-war rural Ireland until years later when I came across some old photographs Renouf had taken in Skibbereen. They showed drab grey houses, one with a poster declaring NO BRITISH WANTED, and unmade-up streets on Fair days, full of cows and ankle deep in dung. Clusters of weathered men wearing bowler hats and smelling of livestock assess the heifers. The children are barefoot, and some of the women wear hooded West Cork cloaks, family heirlooms handed down from mother to daughter. My favourite photo is of a portable dancing platform erected on the north quay of the lough. Couples waltz to the wheeze of a concertina – the women in their best frocks, the men in boots and caps and uneasily encased in suits. I heard that the platform was later vandalised, and they suspected the local priest who disapproved of such unbridled sexual behaviour. He needn't have worried: de Valéra returned from Paris and admitted that 'sex in Ireland is as yet in its infancy'.

'The *Bechstein* sank of course,' John mused. 'That's about all I can remember.'

'Thank goodness for that,' said Jack from the other end of the table where he had been calculating data but also keeping an ear tuned to John's tales. Then he got up and went off to busy himself in the lab next door.

Without Jack, John relaxed, poured us both a gin and orange from a secret supply, and continued the story.

'The students on that first trip loved the place so much that ten of the original twelve decided to return the next year. We

asked Jack, a new young lecturer, to organise a proper expedition. He was the obvious choice because he was nearest to our age and keen on marine biological research.

'It took so long to load our equipment that the train was delayed for ten minutes while the guard chafed at his flag. Jack thought the station master had a henpecked look and was "a possible weakness in future arrangements unless kept under control".'

'How was he going to control him?'

'Haven't the faintest idea. But he certainly fixed him with an icy stare that froze him into immobility. To be fair to the poor guard, we did have a *lot* of gear. It filled two-thirds of the bus from Cork, and we occupied all the rest. There was no room for the impatient queue of regular passengers. Then someone noticed that the bus stop sign read Glengarriff and the Skibbereen bus was standing at a different bay. We'd got on the wrong bus. The inspector quickly changed the plates with those on the real Skibbereen coach, turning out the passengers. The drivers also had to swap, as only the other one knew the way to Skibbereen. There were now two sets of angry, displaced passengers, so the inspector ordered the driver to "start immediately and drive non-stop to Skibbereen". As we pulled out, those left behind started fighting.

'The bus sped away and at all the stops the driver merely slowed down and shouted at the queues, "'Tis all right, there's another one coming . . . sometime." He stopped only to strike deals with shady characters or pop into a bar. A woman with a parrot thrust the cage at a student saying, "Here, mind me bird," and joined the driver in the pub.

'The last leg of the journey, from Skibbereen to the lough, was the most stately, in ancient black limousines supplied by

Donelan's, the local pub and undertaker. I had the sinking feeling we were off to a funeral.'

Having read their first publication from the Lough Ine work, I knew that they were off to sink into the Rapids.

'Why did you decide to begin by studying the Rapids?'

'Well, a problem with field ecology is that natural environments are so complex that it's difficult to unravel what's going on. In those days ecologists monitored the environment in the hope that some factors would correlate with the distribution of the organisms. But, as you know, just because an animal is, say, confined to the top of the shore doesn't mean it *needs* to be dried out twice a day. It may simply be excluded from lower down by some unknown factor. And you can never rule out chance. Even the strongest correlations don't necessarily indicate cause and effect – in the USA the production of pigs mirrors that of pig iron. Uhum.

'Jack's approach was to assume that nature had in effect done all the large-scale ecological "experiments" already and we merely had to study natural systems in which the influence of a single factor was so overwhelming that it must have moulded the community. He reckoned that in the Rapids the current was so strong that it was just such a factor, a perfect demonstration of the effects of water flow on marine creatures.'

'How on earth did you measure the current in those days?'

'Well, for our first attempt we moored a rowing boat at the mouth of the Rapids just inside the lough, but it was caught by the flow, the oars fell overboard and we were sucked down out of control. Jack and I leapt for our lives before the boat capsized and broke up on a rock. We only just made it to shore.

'Eventually we managed to hold a boat stationary in the fierce current on a pair of ropes leading to each bank, where the entire party tried to win a tug of war with the Rapids. A disposable student in the boat held a current meter, like a wind gauge on a long pole, at specified depths. The cup-shaped vanes of the meter whirled in the flow and generated clicks that were counted by a student wearing earphones. We calculated that the volume of water flowing through the Rapids on each tide was equivalent to a Rapids-wide stream twenty-five kilometres long.

'We also mapped the Rapids, and the students became enshrined in survey points: Peggy's Mark, Nita's Rock, Dick's Folly.' John paused for a moment, perhaps remembering long-ago Peggy, Nita and Dick.

'We all stayed in the bungalow, where the garden had been newly planted with exotic cabbage palms and pampas grass. Two ancient local women cooked for us. Their speciality was boiled eggs, as many as we could eat – for breakfast, lunch and dinner. In the evenings they sat by candlelight counting their rosaries and emitting an eerie drone, like bees in a bottle. We augmented our diet with crabs and winkles from the lough, but the "caviar" from sea urchin gonads left something to be desired.'

'And Renouf,' I asked, 'was he still active?'

'Oh yes, and he enjoyed showing us around his lough. It's amazing how this place arouses proprietorial instincts. But he was happy to have us here studying the fauna and was very helpful, although sometimes less help would have been safer. When he was rowing us around Bullock Island, we suddenly had to bail frantically with a hat to give him time to beach the boat before it sank. We had to walk back to bring out another boat to the rescue, but I stepped on the gunwhale and swamped it, so everyone got soaked and our gear shot off down the Rapids. Jack was not amused.'

'Did you never collaborate with Renouf?'

'No. He knew a lot about sponges, but published very little. He had all the sponge samples from our surveys for a couple of years, but never got round to naming a single specimen. Eventually they had to be tactfully retrieved so they could be identified by an expert at the British Museum. The expert gave Jack a copy of his unpublished monograph on sponges, providing it was not shown to Renouf. "I'm sure he wouldn't consciously lift the data," he said, "but it would give his subconscious too much scope."

'The war then intervened, and Jack was drafted into a group at the University of Toronto working on problems of aviation physiology. He rarely talks about it, but I know his speciality was the survival of pilots that had to ditch in the sea.'

Thirty years later I discovered from contacts in Canada that one of his projects was to measure the internal body temperature of men exposed to severe cold. Jack devised a survival suit that looked like a leaky blimp, and then asked to be abandoned in the Atlantic off Nova Scotia in November to put it to the test. The kit they gave him contained fishing gear, so there were plenty of hooks around to puncture his rubber dinghy. He was also given matches should he need to light a fire. He eked out the water supply with sea water and his survival rations included pea-soup powder and chocolate, which he reckoned were 'nauseating'. They must have been pretty bad – Jack was fearless in the face of food. Within months of the trial, the survival suit was issued to bomber crews. It earned him the OBE from King George VI, 'for Ordinary Bloody Effort,' he once boasted.

'I was still a research student studying hormone physiology at Bristol,' John resumed, 'when Jack rang me up in 1946 to suggest a trip to Lough Ine to continue the work we'd begun before the war. So, after bribing a policeman to secure a sailing ticket, off I went to Ireland to reconnoitre. It was a dreary journey from Cork to Skibbereen. The Great Southern Railway staggered through a succession of abandoned stations: Goggin, Knockbue and the long deserted Desert. Only fifteen years later the entire line was closed. The still-extant stations were all painted patriotic green, and one had a small boy flogging peppermints from his mum's tea-tray. When I didn't buy any he stuck out his tongue. I vividly recall that the poster behind him advertised:

B.I.P. Emulsion
Kills vermin in children's heads

'I would have bought some of that all right and used it on him there and then. There were lots of memorable posters,' John mused as he raised his hand to trace the remembered advertisements in the air.

Martin Ltd, Cork.
Superfluous hairs removed for ever
Guaranteed in writing. 2/6 per sitting (50 hairs)

Dance without stockings
Simpson's non-greasy Foot Ointment

'As the train penetrated deeper into West Cork the concerns became more rural.'

Poultrine fowltonic
Invaluable in all cases of
Discoloured heads, Diarrhoea, Drooping etc.

Porkaline pig powders . . .

'You're making this up,' I protested.

John shook his head. 'There is no need to invent Ireland.'

Renouf's labs were available, so they hastily arranged an expedition. How different it was in those days, I thought. Instead of having to devise a justification for the work, and labour for weeks over a grant proposal, all that was needed was a telephone call: 'Come on, chums, let's have an expedition.' And off they went.

'We enlisted as many members of the earlier trips as we could muster. They were a great team,' John continued, and a smile crept across his lips as he remembered them with obvious affection. 'There was wee Ronald Bassindale, Bass, a jolly Yorkshireman and one of my teachers at Bristol. He was brilliant at doing impersonations of all sorts of animals during his lectures to illustrate some point or other. He could also identify almost everything that lived in the sea, and lubricated the proceedings with an endless supply of geriatric jokes, but in those days even the old jokes seemed new. Then there was impulsive Dick Purchon from Cardiff University, who caught prawns in the rock pools and ate them raw to disgust the girls. "It's half an hour before dinner," he said on the first day. "Just enough time to chop down a dead tree, tow it over to the island and have a bonfire." And so we did, and had hot Cornish pasties by firelight. The fun was just to be there with Jack, as part of the team.

'We lived and worked in Renouf's labs. Bass's sister cooked, and six students slaved as research assistants and paid for the privilege. It soon became obvious that the study of the Rapids was a bigger job than we had imagined, so we returned every summer to add more pieces to the jigsaw. In 1950 Jock Sloane, another Bristol lecturer, joined the group. He was a wry, dry Scot, solid and dependable, an ex-Major with a military flair for organisation. You felt that nothing could go wrong if he was there.'

So the team was almost complete, with Jack supplying most of the ideas and discipline. Expeditions are a test of character and patience. I wondered how the respectable conservatism of Jack and Jock had coped with the bawdy exuberance of Bass and John. Years later Michael Sleigh, a student who became another regular, told me that 'Jack made it happen, John made it work, Jock kept the peace and Bass kept everyone amused'.

'Jock told us tales of student days in Glasgow,' said John, 'and Bass of his time on the Gold Coast. And we laughed a lot, Bass hunching his shoulders and almost rolling himself into a ball to chortle. Only Jock kept a straight face through almost any provocation; perhaps being warden of a students' hall of residence helped, or maybe the bottle of whiskey he kept under his camp bed.

'I was in charge of travel, transport, stores and catering and have been ever since. Dozens of crates of equipment and supplies had to be brought in by lorry and then trundled over rough tracks on the cart. British ration books weren't valid in Ireland, so I ordered everyone to bring half a pound of tea with them. After the privations of post-war England, Ireland seemed a land of plenty, overflowing with butter, cream and meat, and lobsters at two shillings each. Here you could eat until turgid, drink yourself silly and smoke yourself to death. Clothing coupons weren't required in Ireland so you could dress well too; for working in the wet I recommended Air Raid Precautions oilskin coat and trousers – and a rare sight we looked.

'By this time I had married Erika,' John added. 'She and her mother had fled from Germany just before the war. Erika worked as technician to an endocrinologist in Bristol and was required to attend classes. When the teacher fell ill with chicken-pox I stepped in, took the class, and later married the pupil. Erika often came as cook on the early expeditions, until the family took up too much of her time.

'The students were scattered in lodgings; two of the girls stayed at the Donovans' farm nearby where the parlour was over-run with chickens. They shared a double bed of extreme concavity so they continually collided mid-mattress, ensuring the fleas were equally distributed. A Flit insecticide sprayer had to be smuggled in. There was lots of food – one of them put on fourteen pounds in a month – washed down with the frothy cow-warm milk we christened "Mrs Donovan's bacteria". The privy outside had fruit wrappers for toilet paper, and on cloudy nights beneath a smoky moon, it would have been a trudge in the dark through tall grass where rats and sheep ticks lurked. As you can imagine, by morning they were bursting for a pee.

'After that we all camped. It started when a student called Rosemary, whose father dealt in ex-army gear, brought a bell tent. At the end of the trip Jack bought it for seven pounds and ten shillings, and got three more. They were heavy camouflage cones, so dark inside that you needed a torch to read the graffiti on the walls. To shield the students from the tales of the sergeant's sexual misadventures, a camp fund was set up to have six new white bell tents specially made for us. We pitched them close to the labs in an unconsecrated graveyard below Templebreedy – Saint Brighid's eighth-century church.'

Saint Brighid was my kind of saint. She could change water into ale and wrote:

> I would like to have a great lake of beer for Christ
>   the King.
> I would like to watch the heavenly family drinking
>   it down through all eternity.

'The field,' John continued, 'was rented from farmer Donovan, who asked three times the going rate. I told him we were only renting the graveyard for a month, not buying it for burial sites. The ruined chapel was once a place of pilgrimage on May Day Eve, and a nearby stone was dimpled by generations of kneeling worshippers. There was also a rune-etched standing stone, and apparently, when drunken sailors threw it into the lough, it was miraculously back in place next morning and the thieves wound up drowned – you know how neatly plotted legends are. I also discovered a beautiful Celtic cross swathed in vegetation.'

'Wasn't it creepy camping in a graveyard?'

'Not really, there were no headstones, and the resentful dead didn't rise from the ground to protest our presence. Well, only once perhaps. The party was sitting around the campfire chatting as it grew dark. We heard footsteps approaching through the bracken and the conversation paused to await the stranger, but nobody appeared. Yet footfalls clearly walked around the fire and then receded back into the bracken. No one saw a thing, but everyone heard . . . something.

'Everyone except Jack, of course. As usual, he had nodded off in the warmth of the fire glow. Jack's a better napper than a kitten. He can fall asleep anywhere, but the slightest touch and he twitches awake, ready to row across the lough. One night he dropped off after dinner still sitting at the table, so everyone crept out taking the candles and leaving him alone in the dark. Next day he didn't even mention it.

'We had everything we needed. There was even a well of sweet water in the field.'

Sweet? I thought. In a graveyard?

'Erika cooked over a turf fire and roasted meat in a bastable perched over the flames. Sediment samples were also dried in an oven over the fire, but kept separate from the food . . . well, most of the time. The utensils became black with soot, and at washing-up time Jack took the largest, dirtiest pot and scrubbed it vigorously to set a good example.

'We all took turns to make breakfast. When it was Bass's stint, he asked everyone how they liked their soft-boiled eggs and wrote the individual timings on the shells in indelible ink. Then he boiled the lot for four minutes and everyone was satisfied.

'In the autumn we filled the larder with wild mushrooms and the blackberries that ripened in the hedgerows. No one went hungry, but Jack always counted the roast potatoes and slices of pudding to ensure he got his fair share. Strict rules governed the allocation of after-dinner biscuits, and whether one cream biscuit equalled two plain was hotly debated. The box circulated clockwise and Jack was known to covet the red jammy one at the centre of the top layer, so everyone conspired to build up his hopes only for them to be dashed just before the tin reached him.'

I could imagine Jack revelling in this pantomime and feigning excessive disappointment, or perhaps filching a jammy one when no one was looking.

'At first, we thought camping might distract us from the scientific work, but it proved to be the making of the trips. The students didn't feel they were just going out into the field to fulfil a requirement of their degree, they came, in Jack's words, "to share in an expedition which was dependent on their own exertions for its success". A spirit sprang up that enriched our work and play.'

'Play?' I repeated in disbelief. 'Jack allowed time for play?'

'Oh yes. There was a marathon hunt for buried "treasure", with the clues growing more complicated each year. We explored the countryside and went night-time fishing for mackerel with two locals, the Bohane brothers. Their philosophy was, "If it's a good time for working, then it's a great time for fishing." Once the boat's engine conked out at sea so one of the kids was lowered over the side to sniff the exhaust and diagnose the problem. It was decided to swap the spark plug for the one discarded the last time this had happened – and it worked. That same night we snagged a seal on the line and headed home at once – it wasn't done to hurt a *selchie*, who, according to legend, could come ashore to live as a mortal.

'We planned the work programme together, and Jack made sure we stuck to it – "Once we have decided how to carry out any particular operation, it is necessary to hold *strictly* to the plan of action" – deviate who dare. We never guessed how much impact our work would have; we were just having a great time and doing some good research as well. The objectives were clear and everyone had a part to play. There was an air of camaraderie and inventiveness. To test the attachment tenacity of seaweed-dwelling snails in fast current, we planted a tough, artificial "kelp" in the Rapids. It was torn to shreds in a single day. When we used a real *Saccorhiza* plant we were able to measure the current velocity at which snails were plucked from their perches by the flow. Jack devised Hobkites – hydrographical observation kites – large-vaned floats that could be sunk to any depth to trace currents below the surface. For surface drift, we used numbered oranges or floating bottles packed with foxglove flowers to make them conspicuous. A boat chased the "float" and a student, precariously upright, held a surveyor's ranging pole immediately over the bottle so that three others on shore could triangulate its position with prismatic compasses. Amazingly, it worked fine.

'To measure the deeper currents we used babies' feeding bottles filled with warm jelly. When tethered below, the bottles

tilted in the flow and the jelly set in the cold water. The angle of
tilt could be translated into current speed and a tiny compass
floating on the jelly gave the direction of flow. When Bass was
asked by a local what on earth was the point of all this, he replied,
"Search me. I get paid to do it."'

'What was the grand plan?'

'There wasn't one to begin with. Jack merely saw the Rapids
as a marvellous system to investigate the effects of water flow on
living organisms. Our first account of the results, *The Ecology of
Lough Ine Rapids with Special Reference to Water Currents*, was published
in 1948, with the authors in alphabetical order to reflect our equal
share in the work. Even Jack never imagined that we would still
be experimenting in the lough sixteen years later.

'We were the only "doctors" within miles, and one night
Jock was called to assist at a difficult birth. All went well and for
years afterwards the family held a birthday ceilidh and invited the
whole party.

'At intervals Renouf, who had no mastery of finance,
demanded a higher rent for the laboratory. "Your students live in

luxury and can afford to pay more," he declared. He also expected redress for missing items that no one had seen, matchwood oars that had been damaged, and chemicals that had been used. Jack jokingly blamed me for the "large quantities of teepol and paraffin which you seem to have been drinking on the quiet". I told him not to worry about such "Renouferies". A list of complaints was just Louis' way of presenting his financial estimates.

'Then, in 1950, out of the blue, Jack announced, "Let us build our own laboratory. That site on the far bank overlooking the Rapids would be ideal."'

The site was Dromadoon, or *Druim a dún* (Ridge of the Fort), for there's a Neolithic mound up there, lost somewhere in the bracken. First they had to get the land – which meant that Jack would get it, because he had private means. Erika and John stayed behind after the expedition departed and visited a local solicitor to find out who owned the land. Before the war it had belonged to Dinny Brien, in his youth the strongest man in the district, but later called 'Mad Brien of the current' because every time he spied a strange boat approaching the Rapids, he rolled great boulders down from the cliffs in case it held invading Vikings. He lived almost entirely on shellfish, and gradually his cottage filled up with discarded seashells. One night out of spite he pinched the buds out of an entire field of broccoli. The field belonged to Carrie Bohane, who was feared because of her terrible temper. She cursed the hand that did the deed and swore that he would be dead within the year. And he was.

'The land was then owned by Mat Burchill,' John said, 'and Jack offered him fifty pounds for a two-acre plot on the steep slopes dipping down to the Rapids. "Why go through a lawyer?" said wily Mat to me. "Let me and himself meet on the site and settle it face to face." Face to distant face as it turned out. Burchill and Jack occupied opposite ends of the field, while Burchill's brother and I alternately conferred with our "clients" and then met mid-field to exchange bids and counter bids. After much shuttling diplomacy, a deal was struck and the men met for the first time to shake hands. "Now that's settled," said Burchill, "I'll bring down all the materials for you for a pound a day." "Done," said Jack, and, true to his word, Burchill carried everything in by horse and cart down a precipitous track, difficult enough to descend on foot.

'Building began the next summer with stints of foundation-laying punctuating the scientific work. Professor Percival, on sabbatical leave from New Zealand, mixed tons of cement, yet said, "It would have been worth coming to Britain just for those wonderful weeks at Lough Ine."

'The Dromadoon lab took almost five years to complete, a bit at a time, with a tyro designer, Jack — the keeper of the spirit level — and the rest of us casting and lifting the heavy concrete lintels into place. The workbench was made from a single piece of mahogany salvaged from a wreck. It was nineteen feet long and two inches thick. We solved constructional problems empirically: when fixing a waste pipe through the wall Jack insisted it should first be whipped with string to give the cement better purchase. Jock thought it was a daft idea, for as soon as the string rotted, the pipe would fall out. So he tied a single, token strand around the pipe. "Did you put string round the pipe?" asked Jack. "Of course," Jock replied truthfully.'

'You built the entire lab yourselves?' I asked.

'Not quite. John Sullivan, a local craftsman, put the roof on when he took his rest from telling wonderful tales. He claimed that the lab was so well constructed that if an IRA bomb blew out all the bricks, it would still stand on the mortar alone. The IRA, of course, wasn't so much of a worry as the winter winds that buffet the coast.

'We invited Renouf to inter beneath the floor an earthenware jar containing a newspaper, reprints of our research papers and other artefacts of 1952, as a time capsule for future archaeologists.'

'So he was still at the lough in the 1950s?'

'Well, only just. A feud had developed between him and Alfred O'Rahilly, the president of University College Cork. They could never have seen eye to eye; O'Rahilly was a Sinn Féiner and Louis staunchly pro-British. It would have been fine if only he hadn't been so outspoken or had chosen a less dangerous opponent. O'Rahilly wouldn't serve on a committee if he wasn't the chairman. Both a Jesuit and a physicist, he used maths to prove that the Shroud of Turin was genuine, and presented a copy of his pamphlet to the Pontiff. It was rumoured that he went to Rome "to give an audience to the Pope". He hovered somewhere between the devil and the Holy See.

'O'Rahilly thrived on conflict; no opponent was too powerful, no cause too slight. Before retiring he somehow ensured that the president's house would never be occupied by his, inevitably inferior, successor. He was omnipotent in his small world, and an autocrat can make a trial out of trivia. He decreed that female students *must* wear stockings, and held regular leg inspections to ensure the rule was obeyed. Renouf's daughter Jean once saw him on all fours in front of a woman, checking her legs. It was like a convent with a mad mother superior. But O'Rahilly was no joke; he was ruthless. When opposed by the warden of the college hostel, I heard that he ordered his cat to be put down, while Louis' punishment was to be locked out of the laboratories he had founded. For a while he camped in the packing case beside the Rapids, then for a couple years in a tiny trailer on the north quay of the lough in sight of his beloved labs – but forbidden to enter, for they weren't really his, they belonged to the university.'

In 1954, John explained, Renouf retired to a house in Cork City with his beloved Nora, their corgis and spaniels, and his collection of Irish stamps. But minor disputes with authority ruffled his retirement. He applied to have his rates reduced when a large depot opened nearby.

'Why should we give you a reduction?' asked the official from the corporation.

'Because,' said Louis, 'the large numbers of undesirable women attracted to the new development have lowered the value of my property.'

'Not at all, quite the contrary,' he was assured. 'They will enhance the value no end.'

Jack visited him a year before he died. When he rang the doorbell a dog's head appeared from every window and the entire house began to bark.

'Make yourself at home,' said Nora warmly. 'Don't mind the dogs.' Easier said than done, as Jack was sharing the sofa with a half-chewed sheep's head. Dinner was less than it might have been because the Champion Spaniel of all Ireland had helped itself to the soup from the pan. Renouf was an enthusiast, and whereas once it had been Lough Ine that had consumed his life, now it was breeding Nora's dogs. Jack was right when he said that Renouf had good intentions, but they never came off. He devoted over forty years of his life to Ireland, his adopted country, yet he remained essentially a British exile. He had charm, generosity and courage.

'For years we'd been far more active at Lough Ine than Cork

University had.' John said. 'Irish students still visited the lough with the new professor, but were forbidden to fraternise with our parties or even with the locals. They must have felt like intruders in their own country.'

It hadn't occurred to me until now, but I hadn't seen a single Irish student here. I was assured that they still made occasional day trips from Cork to the lough, but that was all.

'The trouble is that their facilities here have steadily declined since Renouf's time,' John explained. 'Years ago a winter storm despatched their hut from Rapid's Quay to the Whirlpool Cliff. The Bohane brothers towed it back and replaced it on its foundations, but it was demolished by another gale, drowning all the glassware. A large battery jar that Renouf used as an aquarium tank still serves as a lobster's underwater lodge beside the Rapids. The two remaining huts got worse every winter. By that time we had not only built the Dromadoon lab, but this place at Glannafeen too.'

'Why did you build a second lab?'

'Well, erecting our lab at Dromadoon had peeved Mrs Donovan. "If it was land you were wanting," she said, "why didn't you ask? I have a fine little patch you could have had."

'"Then I'll have it," said Jack, who sold his small yacht and bought the Glannafeen site on the south shore for ninety pounds. On it he built a boat house, and he brought over his beloved brown boat. A scientific balance was installed where, according to Jack, it "would not quake when John clumps around with his big feet".

'Jack suggested to Professor O'Rourke, Renouf's successor at Cork, that they should pool resources, but he wasn't interested. So Jack had a better idea: "Let us build our own mess hut on the Glannafeen site and convert the boat house into a laboratory."

'"Do we need one?" said Jock. "We still have access to Cork's labs."

'"They could fall down any time," Jack countered.

'"But it's taken us five years to build the Dromadoon lab," I said. "Do we really want to start all over again?"

'"I am not suggesting we do it ourselves. I thought I would ask Mr Sullivan to build it."

'"That might be expensive."

'"I would pay for it," said Jack.

'"Well, we're a team. Perhaps we could all chip in a little towards it."

'"There is no need," said Jack. "I can afford it."

'We feared that if Jack owned all the laboratories, it might become *his* show rather than a partnership. But it was his land and his money, so the plan went ahead.'

I detected no hint of resentment in John's voice.

'Jack produced an excellent design,' he continued, 'and it was realised within a year by Mr Sullivan, who was helped by a local farmer's son. The adjoining boat house was converted into a laboratory and we built the concrete quay on the shore. Jack later bought all the coastal land west of Glannafeen to the muddy inlet called the Goleen, and a strip beyond to allow access to the road. So this building became our headquarters and we camped here in the field.'

'What became of the other regulars?'

'Well, Jack fretted about "the social and educational responsibilities of the senior members". I suppose he meant that we should set a good example to the students. Apparently Dick Purchon hadn't done this to Jack's satisfaction, so he tackled him and "settled things satisfactorily". In other words, he laid down the law and Purchon never returned. In the late 1950s, first Bass, then Jock withdrew, probably because they had other commitments, you know how it is. We've all remained friends,' John added, as if to reassure me there had not been a falling out.

'And you're still here.'

'Well, I love the place and the collaboration. Of course, Jack sometimes gets impatient at any perceived inefficiency, although he's good at losing my letters and not finding them until months later. I teased him by writing to express interest in the archaeological excavations in his room and suggested that if any other important objects turned up at the same level, he could use my letter to date them. Uhum.'

As with any friendship there were obviously tensions on occasion – and John relieved them with a joke.

Glannafeen (*Gleann na bhFianna*) means Glen of the Fianna – the warriors of the great chief Fionn MacCumhail. But the legendary Fionn had now been supplanted by a new leader and, in spite of Jack's efforts to consult, I was in no doubt who was in charge.

I had the impression that Jack might have become cantankerous before his time. He had surrounded himself with congenial and gifted people, but then, perhaps inadvertantly, driven them away. Neither Bassindale nor Sloane had ever again published research as good as those early papers from Lough Ine.

I felt sure that the magic of the place still clung to their memories. The magic that Jack and John captured when, in 1950, they had written a popular account of their work:

'Groping by boat in the darkness of a sea cave, and the crunch of the swell on the shingle at its head; fishing at night off the coast, with the Fastnet light flashing on the horizon; the curlews crying over the undisturbed waters of the lough in the early morning, and the soft whirr of the nightjars at evening. And always in our ears, like the echo in a shell, is the subdued roar of the Rapids.'

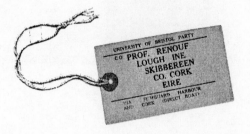

Neighbours

Jack was not given to reminiscing, so the next evening as he was patching his old blue shorts I attempted to get him to talk about things other than crabs and winkles — not by the direct route of asking questions, but by making a provocative statement and awaiting correction.

'Ireland must be the poorest country in Europe,' I asserted.

'Not any more,' Jack countered, 'but when we first came to Lough Ine we were visiting a peasant community where the way of life had changed little for centuries. Many still lived in poverty. To economise on tobacco, a local farmer was given a ration of three matches each day as he left for the fields. Dinner was potatoes with pepper; the father alone might sometimes have a little salt fish. If there was an occasional boiled egg, he would cut off the top for the children to share, and they might also get a spoonful of milk each.'

'Do you remember,' said John, 'the family who relied on cast-offs from the Renoufs and, when there weren't enough boys' clothes to go round, the lads made do with dresses?'

'I do.' Jack replied. 'But in spite of their poverty they were honest. I once lost my watch whilst trekking from Skibbereen. A week later I spotted it on a garden wall. "Ah, we knew you'd be back," said the farmer. "It's a valuable watch."

'Life was short. Even now if you ask a local woman how many children she has, she is likely to reply, "Six and three in the grave." In the old days the adults also died prematurely, often from pneumonia following exposure to wind and rain. The most waterproof clothes a woman owned might be a crocheted shawl and a black felt hat; in the rain dye ran down her face until it was black too. Rather than wear her only boots for a journey, you would see her carrying them over her shoulder until she arrived.

'I heard of a local boy who fell ill and whose mother was too ashamed to call the doctor because the bedclothes were so tattered. She wrote of her plight to relatives in England, and a parcel of new bed linen arrived on the day he died.'

'I was once invited to a local funeral,' said John. 'The corpse was laid out on the table, the clock stopped and the mirror covered. All work ceased in the district and tools were abandoned in fields. First the family and neighbours grieved over the body, then keeners – old women practised at wailing – came in to do the job properly. But it was a poor sad ghost of a wake as I remember. At some, everyone, including the non-smokers, would fill a pipe and swallow quantities of whiskey and potheen – a spirit made from fermented potatoes. There would be music and dancing, even practical jokes if someone didn't keep the lads in check. Later, at the graveside, before the soil was thrown in, the screws of the coffin lid were loosened to ease the deceased's exit in the next world.

'Funerals caused disputes so often that there is a local joke: "You've a black eye – was it a wedding or a funeral?" One cortège halted at a crossroads on its way to the cemetery. At this point, the departed's family tugged the coffin towards the place where his father was buried; the widow's relations tried to turn it towards their plot. Soon the coffin was dropped and fists were flying.

'Jo Bohane, a local woman who lives over the hill beyond the Dromadoon lab, told me tales of sitting beside a roaring fire of furze stumps with the winter wind howling round the cottage, being read to until the voice went hoarse, playing cards by candlelight, and cheating by any lights. And dancing at Sullivan's shed. With so much practice there was not a lad or lass who wasn't a pleasure to partner.

'Christmas brought only meagre gifts, but as presents were rare they seemed wonderful. Jo hung up a secret stocking just in case Santa was richer than her parents – but all she got were holes, because crickets crawled in and ate their way out. They lived in crevices at the back of the fireplace and would chirrup loudly when the fire was ablaze. They were considered lucky.'

'Was there a school nearby?' I asked.

'For the younger ones,' said Jack, 'but for the others it was a barefoot tramp over muddy roads three or four miles in each direction. The children met up at the bridge in the north-west corner of the lough, the boys in long shorts and the girls with snowy pinafores over their dresses. I wonder if they knew that other children congregated nearby in a graveyard for the unbaptised, marked only by a few stones scattered like broken toys. There are at least five children's graveyards within a mile and a half of the lough. In the old days, when country baptisms were infrequent, high infant mortality soon filled a field, and without benefit of baptism, thousands of tiny souls were consigned to limbo for eternity. Catholicism can be heartless,' he added, sadly.

'If you talk to the locals,' said John, who had clearly spent hours doing just that, 'they don't recall cold feet and rain-sodden marches, but searching for birds' nests on the way, or knocking down the autumn chestnuts and hollowing them out with their pen nibs to make whistles. Long days were spent on unsolveable sums and writing essays entitled "A Pinch of Salt" or "A Day in the Life of an Old Shoe". For his essay on "My Cat", Neilly Bohane, Jo's brother, wrote, "My cat had six kittens. We kept two and sent four off to join the navy." He got the belt.

'Sean O'Faolain claimed there were more eccentrics to the square mile around here than in any corner of Ireland, and the residents around the lough contributed their share. One old fella was entirely nocturnal; he fished at night and slept by day. When he gave up going to sea and became bed-bound, he still slept through the day and at night sat up in bed reading the newspaper. But this was nothing compared to the habits of the lapsed gentry.

'Have you seen Lough Ine House, the white building that

nestles among the herons' trees at the north-east corner of the lough?'

John didn't wait for my answer. 'Well, when Louis Renouf first came it was inhabited by Gerald Macaura, a generous local benefactor and a man of many talents, all of them minor, and many titles, all of them spurious. As "Professor" Macaura he toured as a stage hypnotist "curing" smokers while at the same time "putting them through ridiculous antics". "Colonel" Macaura sponsored the Macaura Volunteer Band, supplying it with Besson's Prototype Silver Best Quality instruments and green uniforms trimmed with white and yellow braid. They often paraded on the lawn below Lough Ine House to frighten the herons. He also built a cinema in Skibbereen; the Kinematic was claimed to be "as well-equipped a picturedrome as in any city in Ireland". His finest achievement was Dr Macaura's Pulsoconn or Oscilectron, a device for "reviving the blood circulation", for which he "slogged thirty-seven years to arrive at its present state of perfection". It had the look of a large brass hand drill, like a whisk for ostrich eggs. When the handle was turned, a flywheel whizzed round at frightening speed, the device vibrated alarmingly and threw out an occasional spark for good measure. It was said to cure all manner of complaints and ensure that even the most sluggish circulator would pump with gusto.'

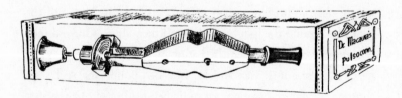

It produced tremors, much as a small earthquake does, and many years later I would come across – purely by chance, you understand – a treatise on *The Technology of Orgasm*. It made the function of the tremors clear. It had long been known to male medics that the root of all feminine disorders was hysteria (literally 'womb disease'), better known as failure to achieve orgasm. They needed help – and help was at hand or, better still, at machine. The Pulsoconn has been identified as a pioneer vibrator. Such a heavy-duty device was surely for the woman who really meant business.

But I knew none of this as John described the sensation created when 'Doctor' Macaura demonstrated the Pulsoconn at City Hall. 'According to the *Cork Examiner*, many of the audience were crushed in the crowd. Hundreds of the devices were sold, until doubts were cast on Macaura's medical qualifications and he fled, leaving a pile of Pulsoconns behind.'

I wonder how many of them were pounced upon by curious matrons?

'And then, in about 1945,' said Jack, taking up the story, 'Captain and Mrs Eyre-Smith moved into Lough Ine House. They were an odd couple. He had been a civil servant in the colonies, and never held any military rank until he retired to Ireland, but the locals thought he should have a title as "he was good at bossing the natives around".

'I was summoned to unblock their well, so I turned up with my diving equipment and climbed in to find the water was only knee-deep. Mrs Eyre-Smith was always collaring me: "I was just chatting to the captain on his favourite subject," she said in her cultured Scot's accent.

'"His favourite subject?" I replied.

'"Radiesthesia," she announced, hoping this would stump me – and it did.

'"Radio anaesthesia?" I ventured, thinking it was the stupor engendered by dull wireless programmes.

'"*Radiesthesia*, silly. Sensitivity to emanations. Simply *everything* has emanations. Flowers have them, so does the grass, even the mattress," she added mysteriously, and gave me a bottle of heady air collected on a Scottish moor.

'On the way back to camp I passed the captain, whose car had broken down. Jug in hand, he was rushing back and forth from the stream filling the radiator, unaware that it had sprung a leak and would never be full. He died shortly afterwards, perhaps from the effort. Somewhere in Heaven, he may still be shuttling to and fro trying to fill the unfillable, or gathering ethereal emanations.'

'And then there was Mrs Banks in the little house on the west shore,' John said with gusto, confident that this would top even Jack's tale of the Eyrey-fairy Smiths. 'It was a very calm evening when Jack and I called to introduce ourselves and we distinctly heard her unroll the hall carpet before she came to the door. It opened to reveal her in a dressing gown, but wearing a large straw sun hat bedecked with flowers and with an avalanche of brilliant red hair beneath. "Would you care for a cup of tea?" she said and ushered us in. The tea would have been fine had she not mistaken salt for sugar, and as her conversation became more animated, she flung her arms about and knocked her hat askew. Her hair tilted with it and almost came off.'

Jack smiled at the memory. 'Her well really was blocked, but this time I despatched our slimmest student, Alan, in a diving suit to unclog it. After the job was done she offered him a warming whiskey from the cocktail cabinet. Unfortunately, she had replaced all the spirits with tap water to discourage drunkenness.'

Disappointed to have forgotten that part of the story, John quickly remembered another. 'Her husband had died some years before and, encouraged by a neighbour, she put a present on the grave, a plug of his favourite tobacco. And it went, just as the neighbour had predicted, so she continued to deposit tobacco regularly, and a crafty spirit never failed to make use of it. Her garden had become overgrown, so she got one of the Bohanes to scythe the grass. He told us he had never encountered such a stony lawn, and that it was only after he had flung the slabs into a heap that he noticed they were inscribed to "My Darling Tiddles" and "Rex – A Faithful Companion".'

'But queerest of all,' countered Jack, now more animated than I had ever seen him, 'was Mrs Hasted and her husband, the major, who lived near Baltimore. She grew vegetables for us and had a soft spot for monkeys. Two big brutes inhabited a large cage in the front garden and flung themselves at the bars, snarling with teeth bared as you walked past. If left free, they dropped on you from the trees.'

'That's right,' John agreed. 'One was brought from Cork by taxi, got loose and amused itself by adjusting the mirror and honking the horn. The vehicle was pulled over by the Garda. "Well, now," said the policeman, staring in disbelief at the ape sitting athwart the driver. "Which one of you monkeys was driving?"'

'On the day we called,' Jack continued unperturbed, 'the monkeys were scampering about the house playing tag and chasing the cat while we tried to carry on a polite conversation.

'"I love it here," said Mrs Hasted above the racket of tumbling brass ornaments and African carvings.

'"So do we,' I agreed.

'"But there are insufficient lighthouses, don't you think?'

'"It is certainly a dangerous coast," I conceded.'

'As the cat flew across the room and leapt on to the dresser,' John added for colour.

Jack was not to be diverted. '"Such wonderful men," she mused. "Braving the elements atop their tall towers with waves crashing all around them." We nodded in agreement, which also enabled us to dodge a falling vase.

'"They must be lonely," she went on, moving to the window to admire the beacon that adorned the headland.'

I had an ominous feeling from John's expression that he was dying to augment the word 'beacon' with something like 'phallus'. Wisely, he said nothing.

'At that moment,' said Jack, deadpan, 'the monkeys caught the cat and bit off its tail.

'"You naughty boys," she chided. "Now look what you've done. Forgive my manners," she said, turning to us. "Would you care for a snack?"

'The detached tail still lay twitching on the carpet when Mrs Hasted returned from the kitchen with tea and sausage rolls.'

Jack then sat back with an air of satisfaction.

Clearly with neighbours like these, an eccentric scientist or two would hardly be noticed. But although the old world of the Anglo-Irish had crumbled away, the marine biologists continued to camp in the past, just as they always had.

'Sounds as if the *Irish* locals were outnumbered,' I observed.

'Well our nearest neighbours, the Donovans of Lake View farm at the head of the Goleen, are Irish all right,' said John. 'In the early days the group was often invited in and offered undrinkable tea. On one occasion a priest arrived and was given a mug of potheen, which he drained as if it were tea. He told of how Saint Patrick had chased the devil from Ireland. "Never to return?" I asked. "Oh he did, thank the Lord. Or I'd be out of a job." And at that moment, from outside came a sickening crash as the priest's ancient car careered into the yard. As usual, he'd forgotten to apply the handbrake. A cockerel and several hens were dead, and others looked destined to join them. The Donovans stressed the value of the departed livestock while the minister mourned his motor car: "After all these years, it is dead then?"

'"Never mind the car, father," said Mrs Donovan, stamping her wellington boot. "You'll have to pay for the birds you've killed. Thirty pounds at least," she ventured, to open the bidding.

'"Lord, no, that's much too much," said the priest.

'"Is it so?" replies Mr D. "But they're dead for certain, and it's your Tin Lizzie that killed them."

'When I told the tale to the grocer in Skibbereen, he said, "Oh yes, he's quite mad. Only last week he ran over himself in the main street. He always leaves the car in neutral, so that if anyone tries to steal it, it won't go." The very same priest, when one of his flock confessed to having ridden a girl from Tralee, replied absentmindedly, "Well, anything's better than walking".'

'Don't you see the Donovans now?' I asked.

'Very rarely,' said John, 'It all changed after old man Donovan died and Sonny became the head of the house. When Jack went to negotiate the purchase of a strip of land round to the Goleen, Mrs Donovan scuttled upstairs and hid, and Sonny conducted the business in flowery, convoluted sentences. As soon as he had made his mark on the legal document, she rushed down distraught, convinced he had sold the entire farm.

'After Mrs Donovan died, Sonny neglected the farm. The buildings crumbled, fields lay fallow and the tracks were taken over by brambles and bracken so that the campsite at Glannafeen was almost cut off. He planted trees at random all around the place because there was a government subsidy. He looks like a shambling wild man now and spends sullen evenings in Skibbereen. If you meet him on the road he turns away until you've passed.'

'When I decided to fence off some of my land,' said Jack, 'Sonny sent a note warning: ANY MORE DAMAGE TO MY LAND AND I WILL PROSECUTE. The solicitor said it was unwise to proceed without Sonny's permission, but he never answered the door or replied to the letters. One day John Bohane – Jo and Neilly's brother, who looks after the laboratories for me when we're not here – spotted him in Skibbereen, but Sonny saw him first and fled, crossing and recrossing the road as John tried to cut him off. They zig-zagged the whole length of Townsend Street and then he ducked into the post office. He emerged immediately from the back door, where John was waiting for him.

'"Ah hello," said Sonny, as if they had met by chance.

'"About this fence," said John Bohane. "Do you have an objection?"

'"Not at all," Sonny replied. And so I erected the fence.'

'Sonny's sister, Kathleen, is even more reclusive,' John claimed. 'Some locals don't even know she exists. I don't think she'd ever left their land until one night when she ran stark naked down the lough-side road claiming that Sonny had "done in" their mother.'

'I saw Kathleen in the fields a couple of days ago,' said Jack. 'As usual, she dived into the hedge as I approached. But I accosted her gently and she slowly emerged and held a reasonable conversation. But there is a look of madness in her. She has wild eyes and haystack hair and is pitifully neglected. It is difficult to believe that in the old days she tended the garden, dried the washing over the gorse bushes and scrubbed pans with sand to make them shine.'

Jack rose to his feet. 'I think it is time for bed. Don't forget to put out the candles,' he said as he trudged off.

John went quiet for a moment and then said, 'We used to spend many an evening with the locals in the early days. We're so busy now that we rarely seem to have the time and it takes time to get to know people. Even then, there are sometimes surprises. Every summer for fifteen years I shared a Guinness with Dinny Salter in his pub at Baltimore.

'"Like many an Irishman I served in the Royal Navy," Dinny told me casually last year.

'"So did my father," I said. "Although I never knew him. He was an engineer and was killed at the end of the Great War when his submarine was sunk in the Baltic on a secret expedition to rescue White Russians. I was ten when they eventually brought home his body to Gosport. I'll never forget the funeral."

'"Neither will I," said Dinny. "I was in charge of the burial party."'

After John went off to bed, the students drifted in to join me in the mess room and I told them ghost stories by candlelight. By the time I approached the climax of the third tale you could have heard blood coagulate. Just as the corpse rapped on the door and the father scrabbled to find the monkey's paw, one of the girls screamed in terror. As I was congratulating myself on a job well done, I turned to see Kathleen Donovan's wild, witch's face pressed to the window pane.

The next day Sonny pinned a threatening note to our gatepost:

KEEP OFF MY LAND AND SEND BACK MY SISTER

The Garda paid him a visit, but there was a rumour that he had a gun so we took no chances. Jack decided that we should never again cross Sonny's fields; we would travel everywhere by boat. It was merely the formalisation of our isolation from the outside world. We were already living in our own special place that Jack had fashioned many years before.

Bohanes

The next day John suggested that we should make a social call on the Bohanes and Jack agreed, so that evening we rowed over to Dromadoon and climbed the steep hill behind the laboratory. John filled me in on the way.

'John and Neilly Bohane have been our friends since the early days. They're both now in their mid-forties and married. John and Philomena are childless, but Neilly and Annie have enough for both families to share. Not that seven children is an exceptional number – the birth rate here in Catholic Ireland is double that in England or the United States. A fifth of married women have seven or more kids, and six thousand have twelve or more.'

'Good grief,' I said from my solitary perch as an only child.

'Over-production is mostly an urban phenomenon,' said John. 'Here in the country marrying late in life is the best means of contraception. Sons age while they wait to take over the farm and bring a woman into the house. The Bohanes solved the problem by sharing their old parents out; Sister Jo, who lives down the lane, looked after their mother, and Neilly cared for their father, Old Con.'

'Old Con?' I asked, perhaps momentarily confusing it with the more familiar expression 'ex-con'.

'Every generation of Bohanes has a Cornelius, and to avoid confusion they're called Old Con, Neilly and Young Corny.' But John was not easily diverted from the matrimonial trail. 'Getting desperate is easier than getting married. Each year dowries are raised as an added inducement and impatient women advertise in the paper's matrimonial column for "a sober man with a view to above". Unfortunately, the men are looking for giddy women with a view to below. A third of all women never marry at all. Although if you'd seen that third, you would know the reason why.'

John guffawed as Jack tried to steer the conversation towards less frivolous waters.

'Most of the scattered houses of Dromadoon were originally thatched and had earthen floors. One had a great sod bank on the side to deflect the winter gales. There were only two rooms, with livestock housed on the left, people on the right. Those that remain have all been slated and floored and some raised to a second storey, but inside they are much the same.'

'Minus the pigs in the parlour,' John added for colour.

The brothers' houses stood fifteen metres apart with a shared well between them. We turned left and there was a spare man with a weatherbeaten face beneath an old cloth cap. It was John Bohane. He had a permanent smile and twinkling eyes. If the farming didn't keep him busy, he could have moonlighted as a leprechaun.

'Ah, Jack, John, it's grand to see you.' He had been sorting a pile of driftwood and glass fishing floats from French and Spanish boats. 'They're like postcards from abroad,' he claimed.

'That shed beside the woodpile,' John Ebling told me, 'is Renouf's hut that washed into the lough years ago, now rescued and re-erected.'

Neilly Bohane heard the conversation and came across to join us. He was a heavier set version of his brother with a deeper, richer voice and a more serious demeanour. There were warm handshakes all round.

John Bohane ushered us into his house with its maritime feel; the doors and banister rails had clearly been rescued from hulks discarded by the sea on this inhospitable coast. A soot-

bottomed kettle, hooked to a chain dangling down the chimney, was forever on the boil. The fire of dark-chocolate turf was fanned by a huge Pierce's Original Blower, a drainpipe coiled like a great prehistoric snail fossilised in iron. When the handle was turned the clatter of the revolving paddles inside drowned all conversation and set tooth fillings on edge. I will never forget the soot-shadowed walls, the aroma of burning peat that clung to my clothes, or the smoke-filtered light seeping from a hurricane lamp hanging from the beams.

Philomena, in her wrap-around pinny, was pressing a shirt with a primus iron, but broke off immediately to offer newly baked soda bread and warm blackcurrant jam, while John Bohane discovered a bottle of Paddy's whiskey. 'You'll be having a little drink I'm sure.'

The table was covered with oil cloth and there were rustic chairs and a high-backed wooden settle beside the fire. Jack pulled a chair forward and the seat came away from its back. He carefully replaced all the radiating spokes of the backrest and hammered them home with his palm.

'There we are, good as new,' he said triumphantly as he lifted the chair, and it fell into sticks at his feet.

'Well,' said John Bohane. 'It has always been a *little* on the loose side.'

There was laughter all round and Jack's embarrassment was dispelled.

'Will I show you some of my jumpers?' Philomena offered.

'Philomena has the niftiest needles in the west,' John Ebling announced. 'One year a student casually said he would love an Aran sweater. He was leaving first thing, two days later, but he took a custom-made sweater with him. Isn't that right?'

'That is true,' said Philomena modestly. Yet in a house full of prize-winning woollies, adventurous V-necks and polychrome cardies, John Bohane stuck with his fuzzy Fair Isle slip-over. John couldn't knit, but he could weave words . . .

The coastal townlands of West Cork were one of the last outposts of Gaelic culture. In 1851 the majority of the locals still thought and prayed in Irish, but the Church decreed that all instruction was to be in 'the new tongue'. English was seen as the hope for the future, Gaelic the badge of the poverty of the past. By 1910, in the school nearest to Lough Ine, all but two of the seventy-five pupils still heard Irish spoken at home, but only one could speak it fluently.

Now Irish clung to the western cliffs by the tips of its ancient fingernails, but here its rhythm and cadences thrived in the West Cork English of John Bohane, more alive than it would ever be in the *Gaeltach* areas, where government subsidies threw lifebelts to a sinking language. He was part of the wonderful Irish tradition of storytellers, for whom, for generations, words and wit had been the weapons of the disarmed.

'Yesterday,' I told him, 'an old guy, a complete stranger, stopped me on the loughside road to assail me with tall tales, rattling on for half an hour and breathing only for emphasis. Unfortunately, I couldn't understand a word he said.'

'Ah, I know the fellow entirely,' said John Bohane lifting and swivelling his cap for punctuation. 'He's the town fool. In fact I think he's the fool for several towns altogether. You should have told him that you'd have to be gone as your house was on fire.'

'I'm sure that would only have reminded him of another story,' I said.

'Then you'd have been best to have knocked him down and run for it,' John said, his eyes at full twinkle.

'I am surprised there haven't been fires,' said Jack. 'The weather has been uncommonly dry.'

'It has so,' John Bohane agreed, 'but in May it was powerful bad weather. Five weeks and every day a pile of rain. We had only one fine day.'

'It was dry all day,' Neilly agreed.

'Well,' John Bohane added, 'except for a shower to keep down the dust.'

Jack went outside with Neilly to look at one of the boats the brothers had been mending.

'I had a go at rowing Jack's brown boat the other day,' I boasted.

'And that would be an experience to be sure,' John Bohane replied. 'I hear it has the mind of a donkey and the will of a woman. But at least it floats. Not like old Professor Renouf's fleet, where every boat was better as a submarine.'

'They certainly weren't all AI at Lloyds,' John Ebling agreed.

'More like ZI I'd say,' John Bohane said. 'He bought a boat from Murphy at Ringarogy. "A tricky man," said Renouf. "You'd as soon get satisfaction from a wall." It was the heaviest boat since the Ark, riddled with rot and repaired with flattened cocoa tins.

'"I shall have Murphy for this," Renouf vowed. "The charlatan has diddled me."

'"Well, tell me," says I. "When a man buys a boat should he best keep his eyes open or keep his eyes shut?"

'"People believe," says he, "that professors have inherited vast amounts of money and are fools into the bargain."

'"Well, I know nothing of the former," I told him, "but I can vouch for the latter." You could pull his leg from here to hell and he'd believe you.'

'Wisely,' said John Ebling, 'Renouf hired Neilly and John to take him out to sea in *their* boat, not one of his own.'

'And what a grand sight he was in his camouflage gas suit,' John Bohane added. 'We landed at Baltimore and Neilly was sent to bring water, but searched all the bars in vain. That day the dredge fetched up a heap of jelly blobs and Professor Renouf knew every one of them by name. On the way back he was for having a picnic at sea, with himself buttering the bread with his finger and handing out tomatoes. We dropped them over the side behind his back instead of eating such new-fangled things. He never noticed the line of little red buoys bobbing off the stern.'

'What do you make of these marine biologists?' I asked.

'They lack no enthusiasm,' John Bohane conceded. 'I remember one year there was a fierce project on limpets. Who would guess there was so much to know in such a little shell? Last year Jack bought lobsters from us. I'm not knowing why he didn't catch them himself. We have only six pots between us, he has forty-seven.'

'He felt it might hurt your feelings if we became fishermen,' John Ebling explained. 'Jack likes to do things in the proper way.'

'That's true,' mused John Bohane. 'He's a very precise man who expects everything to be done *right now*. If I was him I'd never have put my laboratory in Ireland.'

'You realise, Trevor,' Ebling said, 'that John does all the odd jobs for us, perhaps not "right now", but always to Jack's exacting standards. He repairs boats, tends the labs in winter, and he built the large concrete quay on the far side of the Goleen.'

'It was almost done when a county inspector dropped by to admire my work,' continued John Bohane.

'"That's a fine big structure you're building," he says. "You'll be having planning permission I assume?"

'"Planning permission," says I, "what's that? I never heard of such a thing".'

'Jack smoothed it all out,' John Ebling added, 'by assuring the authorities that it was merely the repair of an existing structure – twenty tumbled stones more or less in a line – and we heard no more about it.'

Planning approval should have been difficult to obtain in such a fragile and beautiful place, but John Bohane asserted sadly that a bit of money or influence could make the difficulties evaporate. 'Sure isn't the symbol of Ireland the harp, and it doesn't matter if you can't play it so long as you can pull a few strings.'

Jack and Neilly returned, and after a wonderful warm evening of chat we eventually took our leave. We left carrying a sack of potatoes and a bag of eggs, victuals we always bought from the Bohanes. Neilly accompanied us, as he lived next door.

'Will you show Trevor your parlour?' John Ebling said.

'I will,' said Neilly, and he ushered me into his house and a room that was clearly seldom entered except for dusting. It was dedicated to bowling trophies won by his son.

This form of bowls is almost confined to Cork, and perhaps it's just as well. A twenty-eight-ounce iron ball is hurled along the lanes and the winner is the one who completes the course in the fewest throws. The hurlers sling it underarm prodigious distances, lofting it over the bends to cut the corners, bruising the road and sometimes the spectators too. Fatalities are not unknown, and in 18th-century Ireland it was considered so dangerous that it was banned even when bull-running in the streets was allowed.

'Do people bet on the result?' I asked naively.

'I'd say they do,' said Neilly. 'I've seen wagers of £200 on many a Sunday afternoon.'

'The Irish have always been great hurlers of heavy objects,' John Ebling explained. 'Seven of the first eight gold medals for the Olympic hammer throw went to Irishmen.'

Neilly smiled and added, 'Ach, we've always been grand at swinging the lead, and throwing our money away too.'

De Camp-town la-dies sing dis song

Doo-dah! Doo-dah!

Sundays

I think it must have been John who told me Jack's parents had been Quakers. But his upbringing had failed to give him both God and the Quaker dislike for ritual. Our Sunday evenings came to mimic a religious service — a biscuit, a sherry and a book of ancient tunes.

Every Sunday the Bohanes were invited down. It started well with John Ebling opening a tin of chocolate biscuits and Jack pouring port and Old Nut Brown sherry. Suitably lubricated, the conversation flowed. Neilly was as good a storyteller as his brother with an encyclopaedia of local knowledge but, because of the West Cork lilt and rattling pace, the students deciphered only a fraction of what they said.

After an hour or so the enjoyment came to an abrupt end when Jack signalled we should entertain our guests. It was all right for him — he fell asleep almost immediately and took little further part in the proceedings. Jack had only two speeds – on and off. He could nod off without nodding at all, head erect but eyes shut. Was he really asleep or just quietly listening? I often wondered. I don't think even Jack knew for sure.

Unfortunately, we hadn't consumed enough sherry to improve our voices or dull our ears. Even combined in a bond of mutual sympathy, our reluctant voices were feeble and flat. But then, you can't beat a good dirge.

To make matters worse, song books were distributed.

'De Camptown ladies sing dis song. Doodah, doodah.'

Perhaps they had seemed a good idea in 1950, but in the swinging sixties the songbooks were a mausoleum for mildewed melodies. They were a hangover from when all students found descants a doddle and sang for fun, but that time had long gone. I hadn't seen a song-sheet since I left the Boy Scouts. Indeed Lough Ine was just like being in the scouts; we camped and tied knots and sang songs that were hits when Baden Powell was a lad. If we didn't help old ladies across the road it was merely because there were neither dowagers nor drives available. In retrospect, I think our good deed was to collude to allow one old gentleman to spend his summers in the past.

'We're poor little lambs who have lost their way,
Baa! Baa! Baa!'

Some unfortunate songs were given the Lough Ine treatment. Jack's favourite he called 'Obadahly', the one that began 'Obadahly, obadahly, obadahly Clementine . . .' There was a transient hint of topicality when 'Lloyd George Knew my Father' became 'Keeler knew Profumo, Profumo knew Christine . . .'

Jack snoozed as this tsunami of sound broke around him. We serenaded him with 'Wrap him up in his tarpaulin jacket and say this poor buffer lies low'. We mispronounced 'buffer', and Jack opened a disapproving eye.

Few students could pick up a tune, let alone carry one. One lad could vamp a few chords on his guitar and produced a perky tune while, under the pretence of polishing my shoes, I brushed a brisk rhythm accompaniment. The applause was tumultuous, probably because we had offered a brief respite from the atonal droning.

Thirty years later I was surprised to find the Bohanes recalled these Sunday soirées with affection and a highlight was the night I played the drums on my shoes.

Neilly Bohane's children sat shyly, not knowing what to make of these visitors from another planet, but at a signal they produced penny whistles and played a jaunty jig. Renouf's little granddaughter, Susanna, was staying by the lough for the summer, and she came to sing 'You are my Sunshine' with a wonderful deadpan face and a lisp. And there was nothing to match eight-year-old Corny Bohane singing rebel songs:

> Where the bayonets flash and the rifles crash,
> To the sound of a Thompson gun.

Or even more surprising:

> In the days I went a-courtin',
> I was never tired resortin',
> In the ale house or the playhouse,
> or many a house besides!

And then there was Neilly reciting FitzJames O'Brien's poem 'Lough Ine'. It was written by a solicitor and sounds like it, but not when spoken in Neilly's rich Irish baritone.

> I know a lake where the cool waves break.
> And softly fall on the silver sand . . .

We too tried our hand at verse. Someone recited 'The Highwayman' by Alfred Noyes. We listened intently until, like the ostler in the poem, our eyes became hollows of madness and our hair like mouldy hay. We politely refrained from shouting, 'Stop! We can't stand your delivery!' Years later, I was amazed to discover that it had only twelve verses.

One Sunday the Bohanes brought a detachment of local girl pipers, who wrestled with a tangle of bags and tubes to produce a sound that in the confines of the mess hut was enough to make your ears bleed. It confirmed my belief that bagpipes are best heard outdoors in the distance, the far distance. I believe a true friend is one who can play the bagpipes, but refrains.

The sister of one of the students arrived with Frank, her Californian husband, all beard and sandals and carrying a mandolin on a bandolero of silver dollars.

'Would you play something for us?' I asked.

'Oh, I don't play it,' he drawled, 'I only carry it.'

Instead, he sang 'Brennan on the Moor', under the impression it was 'Brenda on the Move'. He had the loudest, flattest voice I had ever heard, and John leaned over and whispered in my ear, 'I can't quite recall the tune, but I do seem to remember that it had one.'

My efforts recorded the tasks of the week in song. Even Simon and Garfunkel's 'Homeward Bound' was not spared:

> I'm gonna be, downward bound,
> Down! Where the mud is murking,
> Down! Where octopi are squirting,
> Down! Where the worms are lurking
> Silently for me . . .

No icon was too sacred to be lampooned, especially if he was asleep:

> Slumpin' Jack flash I'm aghast, ghast, ghast . . .

If he was awake, Jack's solution was to feed me with chocolate Rolos. Every time I opened my mouth to sing, he popped another one in.

I did, however, learn a song or two, such as 'The Mines', which years later my tiny children insisted upon as a lullaby before they would sink into contented slumber.

> I hope when I'm dead and the ages will roll,
> My body will blacken and turn into coal . . .

My daughter also liked Tom Lehrer's:

> One day when she had nothing to do,
> She chopped her baby brother in two,
> And made him into an Irish stew,
> And invited the neighbours in.

But I suppose that's more understandable.

John Ebling's hoarse, breathless delivery was unique in the annals of musicology. He recited an epic Australian poem concerning a stockman:

He mounted on his bloody horse,
And galloped off of bloody course.
The road was steep and bloody floody.
The creek was deep and bloody muddy.

And 'ere he reached the bloody bank,
The bloody horse beneath him sank.
The stockman's face, a bloody study,
Shouting: Bloody! Bloody! Bloody!

John also croaked such ditties as 'I'm the man that waters the workers' beer' and 'I'm the hole in the elephant's bottom'. But most memorable of all was his rendition of 'The Music Man'.

I am the music man and I come from Fairyland,
And I do play . . .
What do you play?
I play the bagpipes . . .

At this Jack, momentarily awake, held his nose, put back his head and karate-chopped his throat whilst emitting the plaintive squeal of a weasel caught in a threshing machine.

Finally, singing 'Goodnight Ladies', we escorted the Bohanes to the quay and watched them row away into the darkness with their oars smearing the lough with green phosphorescence. We city dwellers had not realised until then how dark the night was and how brilliant the stars.

The air was unusually still and cloudless, so we lingered on the quay in silence for an hour or more to watch a sputnik idle by and shooting stars score the summer sky.

Mailboat

When we were not exploring the shores for specimens, we explored the outer coastline for pleasure. In the early morning, *Naom Ciarán* (Saint Kiernan), the mailboat from Baltimore, sailed into Barloge to pick up our entire party. It anchored in the middle of the creek and Jack, the old man of the sea, ferried us out in a dinghy.

Two green-faced American ladies, having taken the choppy ride from Baltimore, had had enough. They demanded that Jack take them off to stable land 'Right now!'. He obliged and they offered him five shillings, which he accepted. They were put ashore at Barloge Quay and set off up the track, hauling suitcases as big as wardrobes. We warned them it was half a mile uphill to an unfrequented road, but it would have made no difference had it been twenty miles, they were 'not going another inch on that goddamned cork'.

Once we set off on the boat we knew what they meant. The swell sent the sky askew. The boat pitched and rolled and yawed and would willingly have done other things if it could have found a dimension in which to do them. Waves crashed over the bow while Jack, attired in oilskins and sou'wester, sat up front like a comic figurehead and braved the surf. Seeing Jack, the mighty merman, it was difficult to imagine that he had ever suffered from seasickness, but legend (or at least one of Jock Sloane's poems), has it that in the 1950s . . .

> The sea was rough and Jack lay still
> Upon the deck. 'He's very ill,'
> The whisper went, but not to him,
> Because he has his little whim
> about seasickness. He would hate
> us to see him in such awful state.
> But for the journey back we planned
> To help (we hoped) him to withstand,
> The angry ocean's awesome swell,
> We fortified him with a Kwell.

The pill was administered without his knowledge.

The mailboat's deck was painted grey, and here and there the successive layers of paint had scabbed and, like all scabs, it had been picked off to reveal it was yellow beneath. There was nowhere comfortable to sit – the prow was intermittently submerged and the stern shrouded with exhaust fumes. Below decks was dry, but it was like being rattled about in a hot box.

Everyone slumped on deck and hung on like frightened limpets while the mate, with salt-bleached overalls and monkey's hands, skipped around tying and untying ropes, laughing at the disabled passengers. Dan Leonard, in collarless shirt, bright braces and skipper's cap, steered the boat past the Loo buoy, labelled in memory of a Man of War wrecked there in 1696.

We eased out of the harbour, past the Baltimore beacon that sat like a distempered moon rocket on top of the cliff. 'It was erected by Mrs Eversharp,' Jack declared, 'in memory of her late husband, the inventor of the propelling pencil.' The students listened intently, not sure whether a professor would tell them fibs.

The boat moved out into Roaringwater Bay, which roars only quietly unless provoked by a real storm from the south-west. Usually it just grumbles to itself in the gullies: 'The tides aren't what they used to be and today's seaweed isn't a patch on the old stuff.'

We sailed past pearl-laden rivers and a menagerie of islands – Hare Island, Horse Island, The Three Calves – scattered like lost pieces of a jigsaw. These were the 'Hundred Isles of Carbery', 127 if you include all the rocks that dry at low water. The benign inner islands gave way to bleaker treeless islets where unruly waves tried to dislodge the seaweed. Seals had hauled out to enjoy the sun and in the distance porpoises arched above the waves. Basking sharks and hammerheads had been spotted here, and today it was

threshers. They swam alongside the boat with their immense tail fins, two metres long, standing proud of the water. Too exuberant for propulsion, the tail is used to thrash the water while the shark circles to corral its prey and stun it. Thresher sharks are supposed to be harmless to man, but it would be difficult to convince the fisherman who landed one and was decapitated by a single lash of its tail.

Dan's boat had the franchise for taking the mail out to the island of Cape Clear. He steered gingerly into the narrow confines of North Harbour on Cape below the imposing cliffs. It had a quay, a pub, a roadside shrine and a bird observatory. On the wall of the observatory some wag(tail?) had chalked: BED, BREAKFAST AND ALL THE BIRDS YOU CAN HANDLE. The official sign on the lavatory door stated: GENTLEMEN, IF POSSIBLE USE THE BUSHES AT THE SIDE OF THE HOUSE NOT THE ELSAN.* THANK YOU.

The Irish seem to use their toilets sparingly. I saw a public lavatory in West Cork on which the local authority had placed two signs: GENTLEMEN; and, just beneath: NO DUMPING.

The observatory was deserted, for hoopoes were hopping in the hinterland, the lesser grey shrike was impaling beetles on thorns in Galleon Cove, and there were rumours of a rustic bunting. Ecstatic ornithologists had galloped off in all directions.

In 1831, 2000 people had lived on the island of Cape Clear, but now there were only 200 in scattered houses. The locals had once grown oats and potatoes, but now most of the fields had fallen to gorse. Once fishing had paid the rent, for the 'fish were so thick that if you fell in the harbour you'd float on them'. The bounty attracted fishermen from as far away as Spain, and men from

---

* A chemical toilet containing a rich, blue sanitary fluid.

Kinsale came each summer and built fish-curing huts on the island. Most Capers were fishermen in those days, drift netting for herring and long lining for mackerel – but there were no boats there now.

Daniel Donovan, writing in 1896, considered the Cape climate ideal. 'In summer the air is balmy and refreshing, largely impregnated with the ozone. We may therefore consider Cape at this season a regular sanatorium.' Perhaps, but in winter, withering salt winds cut across the naked hills and spray soaks the entire island.

The waves would bring useful flotsam ashore: in one cottage Patrick O'Driscoll had relaxed for years in a deck chair marked *C Deck* which he had plucked from the sea just after the *Lusitania* had been torpedoed in 1915.

The sea had taken far more than it had given, for many a Caper had drowned. When fishing was the trade, almost every house had a wreathed photograph over the mantelshelf in memory of a son or father lost at sea.

Capers were fiercely independent folk who referred to the mainland, only eight miles away, as Ireland. Sean O'Faolain met a man on a nearby island who admitted he was a stranger – he had lived there for only five years. They still spoke Gaelic, and the Government sent schoolchildren to stay with local families during the summer. Only Irish was spoken in the house, but the children played in English. There was a revival of Gaelic literature. Some thought this was Ireland refinding her true voice, but others, like the poet Aidan Mathews in *The Death of Irish*, were unconvinced:

> The tide gone out for good,
> Thirty-one words for seaweed
> Whiten on the foreshore.

Cape Clear was to have fulfilled de Valéra's dream of an Ireland with 'people satisfied with frugal comfort . . . cosy homesteads . . . the romping of sturdy children . . . where firesides would be forums for the wisdom of old age'. But now a lethargy was falling over a community sustained by the rent from boarding students, the dole, and bonuses paid for each child born to a Gaelic-speaking family. It came closer to Sacheverell Sitwell's impression of 'a paradise that is unhappy . . . with the sadness of all things that are a little remote from reality . . . on the edge of the world with nothing beyond it'.

At North Harbour, Jack, the clockwork man, led the way. He always led the way because he had the longest stride and was oblivious to pubs. He intended to traverse the island, three miles at the double, by the shortest route, and was irritated that the road wound without reason. On one side – had he deflected his head – was *Dún an Oir* (the Golden Fort) a 13th-century stronghold of the O'Driscolls, once a powerful local clan. It is

said that one midnight long ago a ghost ship came, and its spectral crew buried gold beside the castle. In the 18th century a soldier garrisoned here dug in vain, and the trench was still discernible.

On the other side we passed the ruins of the telegraph station which, during the American Civil War, had relayed the latest despatches to Europe. Ships would anchor just offshore and jettison the mail in a watertight floating case. Locals raced out to be the first to retrieve it, so that it could be promptly cabled to the mainland, then on to England.

Nearby was a group of standing stones called *Fir Bhréagacha* (False Men) because, it is claimed, they were once furnished with uniforms to fool enemies approaching from the sea into thinking an army awaited them. If enemies were so easily fooled, then surely there was little to fear from them.

Jack had already arrived at the westernmost end of the island, but there was no need to keep up with him – we had the lunch basket. He perched on the very edge of the cliffs, preferring spectacle to security. We picnicked beside him, scoffed Spam sandwiches and imbibed the magnificent scenery. Much as I enjoyed the work at Lough Ine, it was great to have a day away from the camp in a place like this.

The sea thrift flowers had faded from pink to parchment but the birds were still there in abundance. Fulmars circled in silence on the updraft, unable to think of anything to say for once, or perhaps knowing that their effortless flight said it all. Gannets plunged down into the pewter sea at a velocity that would rip the wings from a gull. Gannets and puffins dive to a depth of fifteen metres, and guillemots have been seen ninety metres down, but today they pattered across the surface as if, not content with being able to walk on water, they wanted to tap dance too.

True sea birds are wary of land, but are driven there to nest. The guillemots were lined up shoulder to shoulder on absurdly narrow ledges like pottery statuettes on a shelf. As usual, they had forgotten to make nests and relied on eggs too asymmetrical to roll away. Puffins in orange socks and clown's make-up settled into their tunnels on the turf-topped cliffs, or emerged like bullets to go fishing. Some landed fearlessly at our feet to evaluate the taste of Spam. Soon, the great southward migration would begin and they would be gone. On some days, 2000 fulmars and over 3000 shearwaters an hour have been known to glide overhead.

A few of us walked back to Lough Errul, a lonely freshwater lake fringed with pink pokers of amphibious *Polygonum*. We stripped down to our bathing trunks and shared the shallows with a couple of cows lacking any sense of responsibility at the rear end. Beside the reedy shore were stone basins used long ago for processing flax with the magic water of the lough. It was believed to possess remarkable powers for cleaning crockery or dirty linen, and it was said that, after being soaked for a week, a tarry rag would come up like a new handkerchief. In 1775, an eminent chemist attributed this to the chemical composition of the water. But if, like John and me, he had gone swimming in the lough, he would have been impressed by the appetite and audacity of the resident amphipods, small crustacea who took nips out of our submerged bits. We offered the creatures a soiled plate which they scavenged clean in ten minutes. In the sea too, amphipods are premier-division scavengers – a dead fish stranded in a rock pool becomes a tasty tryst for them and is soon reduced to a fleshless skeleton.

We just had time for a tepid pint to warm us up in Paddy Burke's pub. Jack had left his backpack there for safekeeping before setting out to explore the island. In his absence some joker had filled it with boulders. Jack tried nonchalantly to swing it over his shoulder and plummeted to the floor. He leaped to his feet, pulled out a boulder and made as if to throw it. Everyone in the pub dived for cover, but a dog took a fancy to him and threaded its way through the crowd to bite his ankle.

On the voyage back the mate found the stray dog on board and promptly flung it in to the sea to swim home. The only other passenger was a Californian student who was the epitome of happy hippydom. He was touring Ireland collecting traditional folk songs (some probably written by Americans only a few years before) and *objets trouvés*, not all of which looked as if they had been lost. He spent the voyage in the lotus position humming to himself so loudly that it drowned the noise of the engine.

We also went to Crookhaven, the last town in Ireland. It sits on a long hook of land enclosing an immense harbour. Before the war it was a major port for the pilchard and mackerel fisheries. Once 'you couldn't drag your feet for sailors,' and could walk across the harbour on the decks of ships. Now the bay was deserted and Crookhaven lay abandoned on a forgotten limb of land.

The local pub was run by Mr Nottage, an elfish old man, as Irish as could be, but not Irish at all. He had come from Norwich in 1906 to man the Marconi Wireless and Telegraph station on the headland above the village. Its main purpose was to contact approaching transatlantic liners, as their first communication with the old world. He married a local girl and, when the telegraph station closed, took over the Welcome Inn. John misnamed him 'Dotty Nottage'. Over a Guinness, he shuffled three walnut shells with a pea hidden beneath one of them. The old codger moved them around so slowly I could easily follow the track of the covered pea.

'I bet you it's under this one,' he said.

'I bet it's not,' I answered confidently.

'I say it is.'

'And I say it's not.'

'It is.'

'No, it isn't.'

He was getting exasperated. 'Will you buy me a drink if I'm wrong?'

'I will.'

'All right then, I'm wrong!'

Jacky-long-legs strode off in his seven-league boots with me trying to keep up, past a deserted beach alive with horned poppies and sea holly, then up the long steep road to Brow Head to look for passing whales. On the wind-whipped headland was a ruined signal tower built after the abortive French invasion in 1796. The large force had intended to land at nearby Mizen Head, but

storms and an excessive fear of encountering the English fleet led them to fetch up well to the west. Bad weather and worse leadership resulted in the failure of an expedition that might have succeeded in fermenting a rebellion against the Protestant English. Instead, four ships were sunk and four were captured, and the only French feet that trod on Irish soil were in leg irons.

The mailboat headed homeward by the long route out to the Fastnet Rock. It is the southernmost fragment of land, the 'teardrop of Ireland', for it was the last sight of home for all those who left for America, not so much to seek a new life but to escape the despair of the old one.

The forbidding rock juts out of the deep, dark ocean four-and-a-half miles from shore. It rises 45 metres above the waves, and the great tower of the lighthouse almost doubles its height.

John couldn't resist. 'A magnificent erection,' he declared in a loud voice. Jack was a stranger to vulgarity; unfortunately John relished it. Jack said nothing, but, I imagine, logged a complaint. I wondered what would happen when sufficient complaints had accumulated.

Fastnet looks like Ireland's Alcatraz, and must have seemed so for the lonely men posted there. It was rarely calm enough for a ship to berth, and both supplies and relief crews had to be hauled ashore by breeches buoy.

'Some years ago,' said Jack, 'the whole party landed on the rock and, in exchange for delivering groceries and crates of beer, we were allowed to climb the tower and even sat on the roof, clinging to the ornate lightning conductor. From that eyrie there was a breathtaking view of the torn coastline of drowned valleys with the blue-misted mountains beyond. As darkness fell, the keepers put on a firework display with signal rockets.'

The original Victorian light was built after a hundred souls were lost when an American ship foundered off Cape Clear in 1847. That beacon was a much acclaimed metal structure, but after it fell down the praise subsided. A tower of Cornish granite replaced it in 1903, whose light was measured at 750,000 candelas. Imagine, three quarters of a million candles brushing aside the night. The light is visible over twenty miles out to sea, and even beyond, and over the curvature of the Earth it glows like a hidden fire.

Who would have thought that underwater on this bleak rock legions of crabs and starfish swarmed over technicolor carpets of jewel anemones? Beyond lies the abyss, its bottom silted with a cream-coloured ooze inhabited by sea urchins with slate pencil spines, and a starfish that shivers to pieces when brought up from the depths.

We saw yachts on the Fastnet race rounding the rock, heeling over and plunging in the swell, then swinging back towards Plymouth on the return leg. On the rock platform high above, a light keeper paced to and fro like a caged bear in a zoo. He stopped for a moment to view the race, then resumed his exercise, back and forth, back and forth.

The engine of the mailboat spluttered and failed. While the crew frantically tinkered below, we wallowed in the swell. A flock of gannets flew above, circling and searching, then hurling themselves into the water all around us. It was as if we were being straddled by shell fire.

To our relief the motor growled into life and we headed back towards Baltimore. The boat surged forward, riding the swell in long pulses. We cut through fields of jellyfish, the red-brown umbrellas of the lion's mane, a fierce stinger for sure, but no murderer, whatever Sherlock Holmes might have claimed in

The Case of the Lion's Mane. We warranted an escort of low-flying shags, like dark aircraft hoping to remain undetected by dipping beneath the enemy's radar.

John discovered his trousers had been left behind on the pier and he had to shiver in his shorts. From his pocket he produced the stubs of two tickets to La Scala in Milan and reminisced on the merits of the performance of *La Bohème.* Presumably, on warm nights everyone attends the opera wearing their old camping shorts.

Jack looked on in disbelief. After all these years he had not yet learned that John's stories were to be enjoyed not believed.

When we docked at Baltimore, Jack wanted to make a telephone call. 'Does anyone know the dialling code for Glasgow?' he asked.

'Zero four,' John replied instantly. Jack was impressed.

'It's easy to remember,' John announced in his megaphone whisper, 'because Bed-Eager Girls Love Money: Birmingham is zero two, Edinburgh zero three, Glasgow zero four . . .'

Jack was no longer impressed. He stared at John as if he had caught him handing cash to a tart. I doubt that he memorised the acronym, but he would never forget who told him.

All of us (except, of course, for Jack) were exhausted by the time we walked back to camp. We slept well and woke next morning with salt-laden hair and sun-ripened noses.

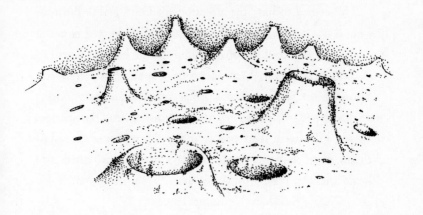

Philosophy

The next day, after I finished collecting more *Saccorhiza* plants, I explored the southern shore of the lough with a student. Beneath a shady overhang, tight-shut madder sea anemones hung down like inflamed udders, and a sponge encrusted the rock.

'Look,' I said enthusiastically, 'it's a crumb-of-bread sponge.'

'Yuck! Sure is a crummy sponge. Looks like an old sock full of holes,' came the reply.

It wasn't the response I had hoped for.

The student began to grumble about the courses he resented having to take at university. 'Biochemistry's bloody awful, all impossible chemical structures and enzyme kinetics. Genetics is just as bad. At school it was OK, just blue eyes and brown eyes, that sort of thing. At uni it's all ratios and calculations. Kills it stone dead for me . . .'

He seemed to have the attention span of a gerbil. But now my mind was turning down the volume while I looked closer at the sponge. How could I convince him that he had misjudged it?

Admittedly sponges lack charisma – 'sponger' is even a term of abuse. In the Skibbereen post office, television licence dodgers were threatened by a poster: TV SPONGERS BEWARE! But sponges modestly hide away their beauty inside where, for a makeshift skeleton, there is a multitude of glass needles fashioned into delicate anchors, tuning forks and stars. They are too small for eyes to see, but too sharp for fingers to miss.

Sponges have a tendency to just sit there and squirt. They have no limpid eyes, no eyes at all in fact, no smile, no fur or feathers or undulating grace and, now we scrub our backs with plastic foam, they're no use at all. Unfortunately it would be another thirty years before the most potent anti-cancer drug we have would be extracted from a deep-sea sponge.

But what about their role in nature as the biggest suckers and blowers on the shore? That was no help. And I had to admit that spongery was an evolutionary cul-de-sac. Sponges never gave rise to anything more advanced than, well . . . sponges.

What about the fact that, if passed through a sieve so that each speck was no bigger than a cluster of only three or four cells, every crumb would grow into an entire sponge, a clone of the parent? Sieve them too fine – into single cells – and the static survivors magically become mobile and aggregate into suitable clusters so that they can go into business as sponge manufacturers. It's a great trick of cartoon resurrection, but perhaps only really of interest to the sponge.

How could I persuade the student that even in the lowliest of creatures there was the wonderful intricacy of detail and yet the elegant simplicity of fundamentals to be found in all living things?

And the ingenuity of design too – even for an animal in the evolutionary basement everything has to be just right, a single element out of place and the organism doesn't work. Then there is the uniformity of processes; the chemistry of respiration that was fizzing away in its cells was exactly the same as in mine.

I moved even closer to the sponge in the hope that it might whisper something in my ear. The crumb of bread must indeed have been stale, for it was greenish grey in colour and smelt faintly of fishy decay. But it wasn't moth-eaten like an old sock, it was cratered. The surface was a moonscape in miniature, graced by tiny volcanoes. Active volcanoes at that, for each emitted a continuous stream of water from the labyrinth of tunnels within.

Being far too large means that there are many terrains we don't explore, many landscapes we never see. Imagine an expedition to ascend one of the volcanoes of a sponge and then to penetrate down into its interior, into the dark catacombs lined with whip-like cilia beating frantically to keep the food-bearing water on the move. The expeditioneers would be doomed, of course, for there was no way they could be belayed so securely as to hold on in such a ferocious flow.

Where my young companion saw only a replica of a dead sock, I could hear the lashing of a million cilia. I turned to him to try to explain, but he was still lashing out at his curriculum:

'. . . and statistics, what sadist invented that? It's worse than a thumb-screw for your brain.' Surely he meant *his* brain.

'Then there's the philosophy of science. What a load of bollocks!'

'You might be right,' I conceded, and turned back to chat to the sponge.

I had no special remit to act as counsel for the defence for sponges, but this tiny encounter taught me that I had a love of all living things that the student would never share. He had an amputation of interest from which I would never suffer.

Being with Jack and John had, I now realised, inspired me. I began to think that I was indeed on my way to becoming a marine biologist.

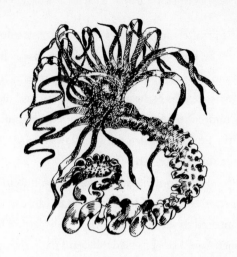

Furbelows

'Best dulse I ever tasted,' I boasted, nibbling at a frond of red alga and stretching my love of nature to its limit.

'How could anybody be interested in seaweeds, let alone eat them?' a pretty blonde student asked.

'I don't know,' I admitted. 'Unless it's because they're works of art and wonders of design and smart enough to time the duration of the night accurately to within fifteen minutes. Or maybe it's because they produce the gels that allow the world's ketchup to flow, keep the head on my beer, the lipstick on your lips, and stop you bleeding to death on the operating table. Then there's the billion-dollar food industry, fire retardants, lace manufacture, and the fascinating case of the never-ending pimento . . . for stuffing olives—'

'All right, all right. Sorry I asked.'

I snacked on little red seaweeds, but studied the big brown

kelps. Their great ecological asset is their size. It allows them to dominate rocky inshore communities. They shade the sun, sieve the waves and provide food and shelter for myriad animals. My mission was to explore these jungles of the sea including the wonderful underwater forest of *Saccorhiza* in the Rapids. If we were to unlock the ecology of the rocky shallows of the coast, the key was to study the kelp forests.

Saccorhiza (sack-root) is the largest kelp found in Europe, up to three-metres tall and as much as a man can carry, but still a midget compared to the giant kelp, one of the largest living things on the planet. *Saccorhiza*'s big blade sits on a flat stalk twisted at the base to act as a shock absorber, with elaborately frilled sides, hence its old name, 'furbelows' (ruffles or flounces). At the base is a warty excrescence the size of a child's head — the 'sack-root' bit.

Every day I had managed to go diving and had collected samples — not just from the Rapids, but from several sites where *Saccorhiza* thrived. Every plant had been weighed and measured. Some had been left in place and labelled with their unique number. They too had been measured so that I could estimate their growth later on. The day-to-day mechanics of research are routine, and the essence of scientific inquiry is measurement. To describe is one thing, but to quantify is far better.

My collections showed that it was a botanical chameleon. In the Rapids the *Saccorhiza* plants were long, thin and tough, with their blades split into as many as thirty straps; whereas those in Renouf's Bay were short and broad with no straps at all, and were so fragile that they ripped under their own weight when lifted from the water. Could these really be simple environmental variants growing only two metres apart? Jack was forever moving animals into new sites to see how they fared, so I decided to transplant my kelps. Sure enough, incipient Rapids plants grew into typical calm-water types when moved into Renouf's Bay.

Plants are wonderfully indeterminate. Their shape is not fixed from the outset, the way it is with animals. A lamb, whether grown in a barn or a hillside, is destined to grow into a sheep-shaped adult; but the shape of trees is sculpted by the wind, that of seaweeds, by the waves.

When exposed to strong currents, *Saccorhiza* plants become elongated, tough and pliant, and the numerous straps allow the big blade to fold and conform to the flow. When a storm wave hits the shore it has the same effect as a wind speed of over 900 miles per hour (a mere 80mph constitutes a hurricane). Little limpets cower in their shells and no oak tree, however mighty, could withstand such forces. The secret of surviving the surf is not to confront it, but to yield to it.

The same strategy was necessary with my supervisor. Like Jack, Dr Burrows had generosity, courage and occasionally humour, but on the whole I would rather she'd kept them in Liverpool. Unfortunately, she was unsure how long I could be trusted to behave when I was out of her sight, so she was arriving in camp today.

Research students, like me, were apprenticed to an established academic who became their supervisor, mentor and thorn in the flesh. It was their task to keep you at it, criticise your work and ensure that you published in the most esteemed scientific journals to bring kudos to you both. Dr Burrows was an example of someone who had done excellent, and sometimes ground-breaking, research, but had published in lesser-quality journals, so her work had been largely overlooked.

There were only two types of supervisor: those who were

always peering over your shoulder and tut-tutting, and those who were nowhere to be seen until you weren't looking. Both were an impediment to self-esteem, playing football and going out with girls — so type one was never sought out and type two was best avoided. Most research students managed pretty well on their own, although sometimes the bulk of the research had to be crammed into the panic-riven final year of the project.

As I was still in the first year of my three-year studentship, I had plenty of time, although I was an enthusiastic if indisciplined workman lacking the tools to do the job. I felt that nature was eager to be explained and its mysteries could be unravelled . . . somehow. What I needed was a kick up the backside.

I steeled myself for Dr Burrows' arrival. She was short and plumpish, with fly-away glasses and hair cropped by the grim reaper. She was a warm but private person, reserved but prickly, and formidable when roused. She always kept a motherly eye on the female students and a stern watch on the males. I guess I was her most troublesome male.

Her name was Elsie, but friends called her Bunny. For me to call her that would have been unthinkable. Colleagues were 'Professor Burges' or 'Dr Dixon' (at least in front of the students), technicians were 'Tony' or 'Stephen'; female students were 'Miss Smith'; and I was simply 'Norton'. But this secure arrangement was to be shattered at Lough Ine.

Jack, who had rowed across the lough to pick up Dr Burrows from the roadside, brought her into the mess room and made a public announcement to the students.

'This is Bunny Burrows from Liverpool University.'

'Hi, Bunny!' the students called in unison, and a blush suffused her face. It was as if the Pope had been introduced as 'Toots'.

It was not that Bunny had illusions of grandeur – far from it – it was just that she felt more secure when everyone knew their place. Familiarity invited students to take liberties, and who knew where it might end?

'Bunny is here to keep an eye on Trevor,' Jack continued.

'He hasn't been misbehaving?' she said in alarm.

'Of course not,' Jack assured her. 'Apart from trying to wreck my brown boat and encouraging John to manufacture even worse jokes than usual, he has been a great asset.'

'Bunny!' John bellowed as he came through the door, rushed up and gave her a hug. 'Terrible fellow this Trevor,' he boomed.

Bunny lowered her voice but I caught the words: 'He is a very nice lad,' she whispered, 'but you have to slap him down occasionally.'

'We can do that with pleasure,' said John.

Then she turned to me. 'Well, how many plants have you processed so far, Norton?'

I saw the puzzlement pass across the students' faces. Norton? Who was Norton?

'Over a hundred and fifty, Dr Burrows,' I replied.

'Good. You made sure that the labels didn't cut into the tissue and weaken the plants?'

'Of course, Dr Burrows.'

'Well done, Norton. You must have been working after all. That's not like you. I suppose there are few opportunities to play football here.'

'Very few, Dr Burrows.'

Some of the onlookers began to giggle. Two more 'Nortons' and a 'Dr Burrows' later, several students collapsed in hysterics.

Bunny had been to Lough Ine before, in the 1950s, and clearly got on well with Jack and John. But from now on she would monopolise my time as she kept me busy with my own work and got me to help with hers. Even so, I saw little of her – well, little of her face but quite a lot of her backside. She was tracking *Codium*, a green rubber glove of a seaweed that had invaded the British Isles and was spreading. I spent several days rowing round the lough as she leaned over, the stern of the boat pinching a tip from every plant we passed. For the deeper plants we swapped places and I leaned over wearing a mask and snorkel. After my first attempt I straightened up, automatically clearing my snorkel with a sharp blast of air and squirting a jet of water straight in her face. She was soaked. Of course I apologised profusely, then resumed the task. I stretched to grab a plant that was almost out of reach and came up triumphantly with it in my hand. And squirted her full in the face again.

A year or so earlier Bunny had had a hip joint replaced, which had restricted her mobility, so often I was despatched to collect at the water's edge. Unfortunately the nesting terns took exception to this and would launch a splash or crash attack, as lime dumpers or diving darts, aiming their lances straight for my head but swooping aside at the very last moment.

'That was very exciting,' Bunny said. 'When we get back I will buy you a beer. After all,' she said with tears in her eyes, 'one good tern deserves another.' Then she collapsed, helpless with laughter.

The weeks at Lough Ine had flown, and it was time to leave. I was sorry to say goodbye to Jack and John.

'But you will back next year?' said Jack.

'Well, it all depends . . .'

'But you *must* come,' John insisted.

Bunny interceded. 'If you are brave enough to invite him,

then I will send him. It will allow me to recuperate during the summer.'

I made my farewells to the students and Jack rowed us across the lough with the solemnity of a ferryman crossing the Styx. A taxi awaited us, supplied by Ernie Donelan, the undertaker.

The Customs official in Dublin looked as stern as only Customs officials and public executioners can. 'Anything to declare?'

'Nothing,' said Bunny, tartly.

'Nothing at all, ma'am?'

'I said nothing, didn't I?'

So we had to open our bags.

'Well, well, what have we here?' said the Customs man, withdrawing a plastic bag full of greyish-white powder from my rucksack.

It was a fair cop. 'It's dried sediment from the bottom of a lake,' I confessed.

'Is that so, sonny? Well, let's have a little look, shall we?' And, just like in the movies, he dipped in his finger and put it to his lips, and I was tempted to ask just how he could recognise the taste of a proscribed substance.

He glared at his damp finger and then at me, and said, 'Best sediment I ever tasted.'

Blarney

The following spring I had to return to Lough Ine to measure the plants labelled the previous summer. Bunny decided to take her new female research student, Win Price, with us, so that she could collect specimens for her project.

Win was tall, slim and elegant — a bit out of my class perhaps. When she'd first become a postgraduate, a fellow researcher had said, 'Have you seen Bunny's new student? She's a real cracker.'

'Oh,' I replied, 'do you think so?'

Little did I know that four years later we would be married.

Bunny must take some of the blame. On the ferry to Dublin, her cabin was on deck A up at the bow, while Win and I were on deck B near to the stern, in cabins with a communicating door.

We drove down from Dublin behind a truck, with bricks of peat falling off at intervals and bouncing clear over the car. We passed a sign for Tipperary. 'I believe it's a long way to Tipperary,' Win quipped.

'No, not very far,' Bunny assured her.

After a detour we got lost and I asked a farmer the way.

'Is this the road to Cork?'

'Cork?' he said, removing his cap and scratching his head.

'The city of Cork,' I repeated.

'Cork?' he said again deep in thought. Then, suddenly, 'Ah, Caark!' as he deciphered my accent.

He pointed, and around the next bend the city lay before us.

There was time for a brief visit to Blarney Castle, but Bunny wouldn't let me kiss the stone. 'No fear!' she said from the heart.

In Skibbereen Bunny lodged at the West Cork Hotel and we were at the Eldon. 'It is because I booked my tickets through a travel agent and yours with American Express,' Bunny explained. Maybe it was true, but she was a very private person and the thought of having to make conversation over the breakfast table with two students probably filled her with dread. None the less, she must surely have realised she was exposing Miss Price to grave risks, and that even the creakiest of floorboards in the corridor outside her room would not save her.

Our hotel turned out to be better provisioned than Bunny's and before we could share a sherry in the evenings she had to wait for us to finish devouring mountains of scampi and doorsteps of chocolate gateau. I later discovered that she had paid for the trip out of her own pocket.

We had picnic lunches for which Bunny would buy desert-dry soda bread and slabs of mousetrap cheese. As we risked our teeth and did our best to chew through it, she said, 'You young people don't know what real food tastes like.'

Yes we did – door wedges and coconut matting.

✻

My plants had survived the winter with their labels intact, and I was able to measure their growth rates. Everywhere else I had studied *Saccorhiza* it was an annual plant that arose in the spring, grew through the summer, reproduced in autumn and then progressively decayed back over winter to make way for the new crop. But the forest in the Rapids was uniquely perennial.

It taught me that every time you think you have discovered a generalisation, you are about to find an exception to it. I found this stimulating, not irritating, for the phenomena that *don't* fit the pattern are usually the most interesting of all. Scientists are people who truly understand that the exception proves the rule, in the sense that it forces them to *test* its validity in the light of data that challenge it.

In the other populations I had studied, a few plants were out of step with the others. They started life later in the year, missed the reproductive boat of that autumn, and overwintered intact to become fertile in the following spring. Only then did they decay. In the Rapids, because the forest was so dense, its fronds excluded the light and new plants sprang up willy-nilly throughout the year whenever a gap appeared in the canopy above. So, although individuals still lasted for only a year or so, there was never a time when they all died back together and the forest noticeably thinned.

The growth rates were high; when it grew it *really* grew, for it had to to produce a large plant in a short time. Jack's pioneering dives off Plymouth in the 1930s had shown that *Saccorhiza* could be abundant one year and absent the next, only to return the year after that. It was an opportunist whose rapid growth allowed it to pop up whenever there was a space. But here in the Rapids it ruled supreme and excluded all rivals.

*

That August I returned to Lough Ine. Bunny had gone ahead in July and I contrived my trip to ensure the minimum of overlap.

She spotted me across the lough, arriving by boat. 'That's Norton. I'm sure it's him. He's not supposed to be here until the twelfth.'

'Never mind,' said John. 'Trevor's welcome whenever he arrives.'

'That's just like the boy,' she said huffily. 'We agreed the twelfth.'

'What is the date today?'

It was the twelfth.

It made no difference. She was still cross with me for making her look foolish by arriving on the right day. So I kept out of her way by taking a sickle and hacking back the bracken that encroached upon the campsite each year, hemming us in against the outside world. Bracken is so toxic that a few mouthfuls can lay low a cow. On warm nights its aroma drifted across the campsite from Sonny Donovan's neglected fields and we woke to the scent of cyanide. 'You cannot beat fresh air,' said Jack, filling his lungs with the stuff.

As Bunny was still quietly simmering when I finished bracken clearing, I went out again and cut a path to the Goleen so that, instead of having to row groceries all the way from the postbox, we needed only to ship them across the narrow Goleen then carry them speedily to camp on foot. Jack tactfully named the new trail 'Trevor's furthest south', rather than 'Trevor's furthest from Bunny'.

A few days later, Bunny departed towards the east.

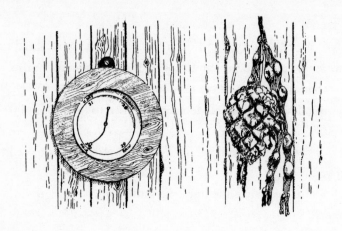

Duet

'John!'

'Yes, Jack.'

'What about tomorrow's programme?'

This conversation took place every evening just as everyone began to think about going to bed. It usually took an hour's debate to arrive at a plan of campaign. Should we risk a day riding the waves at Carrigathorna, or settle for the shelter of the south shore?

Actually, it all depended on the weather. Without access to forecasts (Jack had banned radios from the camp) they relied on weather lore – pine cones, migrating fulmars, red skies at night, etc. – and, of course, Jack was a lore unto himself. Luckily we had a barometer, but not for long. Jack liked to tap it before taking a reading, in case it had fallen asleep. Unfortunately, an over-vigorous tap cracked the glass and bent the arrow. Great faith was still invested in its prognostications, although it was uncertain whether the tip of the arrow or an extension of its original line of flight led to the prediction. This made a difference between 'dry' and 'rain', so we mostly settled for 'changeable'. This being south-west Ireland, we were usually right. Forecasting became much easier when the glass was over-tapped again and the barometer crashed to the floor.

'Is the glass falling?' John asked, pursing his lips.

I was not sure that Jack enjoyed John poking fun at him, but John's mission in life was to find fun in almost everything, even in places where few others cared to look. It often crossed my mind that had he and Jack not been collaborators, they might not have become friends.

Jack retrieved the barometer and was ceremoniously replacing it on the wall when its spring gave a last *boing* and expired.

I picked up a little weather wisdom. I seem to remember that wind from the north-west was good, westerlies, although unpredictable, often erred on the favourable side, and northerlies were either good or bad as the mood took them. Southerlies were always bad, unless it was very anti-cyclonic, in which case you ignored the slight breeze from the south. Fine weather first thing in the morning was also a very bad sign — unless, of course, it remained sunny for the rest of the day.

My solution was to wear shorts and a T-shirt all the time and at the first sign of rain to take off as much as decency would permit and sit on my clothes to keep them dry. I reasoned that it was easier to dry my skin than my meagre wardrobe. As a result, in wet summers like this one, I would spend a great deal of time undressed, and got much browner than in sunny ones.

I had not only been researching *Saccorhiza* over the past year; I had also been enquiring about Jack and John. My most important discovery was that academics were as willing to gossip as to instruct.

Although Jack and John were both zoologists, neither was primarily a marine biologist. Both had loved the sea and its inhabitants since youth. Jack, when six years old, had been taken on holiday to Cornwall. Holding his father's hand, he'd gazed into the tide pools and roamed the fields searching for grass-hoppers and butterflies. Later that same week his father failed to return from a walk. He had collapsed from a heart attack and never recovered. That first fateful holiday had forged Jack's life: he had lost his father and found an enduring love of nature.

Later he went to Cambridge to study classics – for in those days only weak students did science – but changed to biology as soon as he arrived. He became probably the most distinguished experimental zoologist to study the protozoa, those tiny aquatic animalcules that swim in an almost invisible world where, for them, water is as thick as treacle and a hair's width away the conditions can be terrifyingly different. The shapeless *Amoeba*, and the gliding *Paramecium*, which looks like a footprint with a frayed perimeter of cilia, both live in fresh water and as they are not waterproof they tend to flood. To ensure they don't absorb too much and burst, their life is a constant toil of bailing. They have tiny pulsating bilge pumps to expel excess water. Jack revealed how such vulnerable single-celled organisms adjusted their water balance and coped with this stressful environment.

He also found that microscopic animals were useful for studying how cell processes are affected by high pressures such as those endured in the ocean depths. In laboratory experiments the organisms often became befuddled and forgot how to grow or reproduce or get into reverse.

According to legend, the young John Ebling 'was so excited by aquatic life that he couldn't pass water', but he went on to delve into the interior of mammals and became an expert on the relationships between hormones and skin gland secretions. His PhD thesis revealed that hair-growth in rats was under hormonal control. This caused barely a ripple on the scientific pond until another worker showed the same thing in man. As a diversion, while working alone on a lonely island in the Hebrides, he weighed the daily clippings from his electric razor and found that energetic hair growth followed masturbation and preceded seeing his fiancée. Obviously, even the thought of her was enough to make his hormones surge and his hair stand on end. It is the only scientific paper I know that was published anonymously, although everyone thinks they know who the author was.

John was later fêted by the cosmetic and pharmaceutical industries, who valued his opinion on human skin. *Which?* magazine employed him to assess faces that on one side only had been treated with anti-wrinkle cream. What a half-smooth dowager looks like is beyond my imagination.

I would later see him on television discussing the relative merits of such diverse 'cures' for baldness as chicken manure and 'virgin's humours' — a medieval remedy that was no longer available in the 1960s. He wrote articles entitled 'The mythological evolution of nudity' and 'Human hair and blushing', and edited *The Future of Man*, as well as the definitive *Textbook of Dermatology* — two volumes, each as thick as a brick and very expensive. Every medical student worldwide simply had to have it and John raked in the royalties.

Jack and he were both christened John and were true intellectuals interested and informed about the world around them. And they both loved to travel.

'I prefer Milos to Naxos, don't you?'

'Is it Milos that has that delightful little Turkish delight shop on the harbour front?'

'Isn't that Andros?'

'Just consider,' John began, 'the enormous influence of Turkish delight on the cultural development of the eastern Mediterranean . . .'

Yet we all knew that the best Turkish delight we'd ever tasted was Hadji Bey's, softer than jelly and dusty with icing sugar, in tins that had always been old-fashioned. It was made in a little shop in Cork.

The differences between them were far more numerous and striking than the similarities. They worked at different speeds – the rattling rhythm of Jack's rowing, *topocher ker plunk, topocher ker plunk*, the languid slap of John's oars. Jack kept lean by jogging through the lab to demonstrate he was busy; John sometimes broke in to a stroll and was putting on weight. They were different in temperament as well as metabolic rate – Jack fussed over the scientific work and John ensured that the knives on the table pointed in the right direction.

John – Attila the pun – had an exuberant sense of humour, loved jokes, the more ribald the better, and laughed like a lavatory flush, while Jack rarely let himself go and *never* smiled at the risqué. Laughing didn't come naturally to him, but when he did, it was as if he had learned it from a book. For politeness, he uttered a restrained guffaw that really did sound like 'Ha! Ha! Ha!' Very occasionally, if you could take him by surprise, he would forget his laughing lessons and, without making a sound, tilt his head backwards until tears streamed down his face.

Jack was the outdoor man who would readily wade into the frigid sea for samples, but I never saw him submerge for pleasure.

John, on the other hand, was rarely wetted by work but often went for an evening swim in the lough, wearing something slightly smaller than the briefest of swimming trunks. He rode higher in the water than a rubber ball and there were few stranger sights than John, his chest protruding clear of the surface as if he were standing on the bottom, slowly breast stroking towards the Goleen while singing 'Jerusalem' at the top of his voice.

John was an inveterate name-dropper, but Jack indulged rarely. However, when one of the Bohanes suggested that 'the Queen must be a very stuck-up woman', even Jack couldn't resist retorting, 'She seemed very nice when we chatted at her garden party last weekend.'

Jack railed against the inefficiency of the local firms, while John moaned about the lack of organisation nearer to home. John took refuge in the occasional gin and orange and kept a bottle ready mixed in the pantry, disguised as the orange cordial Kia-Ora. Jack collaborated by pretending not to notice, so as not to spoil the secret.

But they both loved Lough Ine. For John, it was a summer visit to the seaside and a chance to see old friends. He packed a camera and a good book and pale-blue pyjamas piped in navy. For Jack, it was a time to play Swallows and Amazons and, of course, do some serious ecology. He brought only the minimum of personal effects as befits an expedition. He left his gear behind at base camp each year, including traditional pyjamas striped like a deck chair and with flies torn right around to the back seam so that the legs would have departed in different directions had they not been tethered together by the cord. John's photographs were of people and cloud-crested hills and carts going to the creamery. Jack recorded the collecting sites and the work in hand.

There was the danger that John might be seen to be slacking

– in Jack's view perhaps the greatest sin of all. Fortunately, they were bound together by their mutual interest in the lough, for, if the expeditions were to succeed, they needed each other, perhaps more than they knew.

Jack had enjoyed being a founding professor at the University of East Anglia, designing the curriculum and drafting the specifications for the new buildings. Unfortunately, he was head of a department where democracy reigned, and he was not adept at the management of people. He was better suited to the old days when the professor ruled. Once upon a time only the professor was paid and if he needed teaching assistants, he funded them from his own salary.

'I can remember when universities were places of learning,' Jack said wistfully.

'They were places of learning,' John claimed, 'only because students arrived with a little knowledge, left with rather less, and it just tended to accumulate.'

'If only the government would stop interfering, we would have a sporting chance.'

'One is only given a sporting chance when they know you have no chance at all. Uhum!'

'Even the students are not what they used to be.'

'You've forgotten Zero Smith at Bristol.'

'Scored zero in every test he sat.'

'That's him.' John beamed, assuming his pleased-with-himself expression. 'Couldn't tell a proton from a crouton.'

Throughout the conversation Jack stood stiffly upright, his arms by his side. He bobbed up and down on the spot when Ha Ha Ha-ing at John's jokes. John was more animated, often lifting up his hands, fingers spread like startled starfish, palms inwards, as if clamping an invisible loaf.

I knew Jack only at Lough Ine – Jack gone feral – but I always suspected that he of the cloven pyjamas and moth-mangled sweaters was the real Jack. 'Everyone needs a bolthole,' he told me. 'A gap in the hedge.'

Lough Ine for him was life with the splinters removed. Back home at Norwich, the students told me, he could be prickly and unpredictable. If you knocked on his office door there might be a warm 'Come in' or a strident 'Go away!' in response.

The research students fared best. He was genuinely interested in their work and generous with his time and advice. But there was continual conflict with his colleagues. Jack loved to put the other point of view, even if he didn't agree with it, striving to be reasonable to the point of unreasonableness. At staff meetings he regularly threatened to resign.

At work his pleasures were confined to having all the laboratory windows open to let the winter in, or blasting them out when his experiments on the effects of pressure went wrong, and, of course, overtaking much younger colleagues on the stairs. On warm spring days he would don shorts and hold classes out of doors and pretend he was at Lough Ine.

His garden at home was his refuge. In winter, come shower or shine, he would walk to the university and back for exercise, carrying his papers in an old shopping bag. In summer he drove, so that he could speed back to his irises.

When digging a garden pond, he kept tatty field clothes in his office so that he could change before rushing home at lunch-time and shifting a few more spadefuls of earth. On one occasion a rep arrived and insisted on taking the head of department out to lunch, but his jaw dropped when out of the professor's office stepped an 'ill-dressed scarecrow with trousers held up with baler twine'. The rep rallied and proffered the invitation.

'Golly, no,' said Jack. 'I'm off to have lunch with my wife.'

Lough Ine was the greatest hideaway of all, a kingdom where he truly ruled. Where there were no radios, no telephones and, best of all, no meetings. Unless, of course, he called one.

'John!'

'Yes, Jack.'

'What about tomorrow's programme?'

Skibbereen

Without a telephone or a vehicle, our link with the outside world and the location of delivered stores was a large rabbit hutch of a postbox at the roadside above the Goleen quay. The mail was brought by Willy Neal and, of course, we called him Willy Nilly the postman. We were at the furthest limits of his round, a long bicycle ride on a hot day and even longer on a wet one, so we often left a bottle of Guinness in the box to raise his enthusiasm for the next trip.

I rarely got a letter, but that summer I was arranging a scientific meeting in Ireland and received a telegram from a fellow marine biologist, Professor Máirín de Valéra, the president's daughter. She was well known in Ireland, as she often accompanied her widowed father on state occasions. A cable from such a lofty source would surely have caused a stir. 'At least a pipe-band accompaniment for the postman,' claimed

John. But it was signed simply, Maírín, so no one knew.

Beside the postbox was an ancient black sit-up-and-beg bicycle with tyres like a tractor and a padlock and chain fit to moor the *Queen Mary*. At least once a week, Susan, our cook and the wife of one of Jack's colleagues at East Anglia, trundled into Skibbereen on her 'Prim Nellie' velocipede to order provisions. It was only four or five miles, but it was hard going with a stiff breeze in her face and uphill most of the way and, inexplicably, just the same coming back. Jack was teased because he had never learned to ride, and ribbed even more when the bike he had owned for twenty years was stolen and he couldn't identify it.

One day Susan met a travelling fair, and the driver at the head of the convoy flagged her down. 'What day is it?' he asked.

'June the twelfth.'

'No, what day of the week?'

'It's Wednesday.'

'Jaysus, Mike, turn around. We should be in Bantry.'

Skibbereen was less than picturesque in those days, before the residents bothered much with keeping up appearances and every garden in Ireland grew a speckled poplar and a bush of pink mallow.

A road sign that had stated, SKIBBEREEN WELCOMES CAREFUL DRIVERS, had been altered to SKIBBEREEN WELCOMES CARE*LESS* RIVERS.

The town lay beside the River Ilen, and sometimes beneath it. In the worst flood the water rose to the tops of the doors and residents escaped through the roofs. A greyhound was seen floating down Townsend Street on a table, followed by a cat in a tin bath.

On dry days, cars and carts parked wherever they had a mind and the *Garda* strolled around chatting to the drivers.

The town had a fine array of businesses including Mrs Lavumba's Coffee Shop and The Star Ball Retainer Company of Ireland – 'Exporters of High Precision Ball Retainers'. The produce on offer wasn't always obvious from the Victorian painted shop signs. The Saint Francis Store sold mostly footballs and sporting magazines, but a passing girl crossed herself just in case, and the butcher's window displayed more pelargoniums than meat. On an adjacent wall a poster announced: PURCHASE OF OLD AND UNECONOMIC COWS. An enigmatic display offered a heap of Tizer bottles, a Panama hat, assorted cleaning compounds, a bowl of corn flakes and framed picture of the Pope – a metallic-sheen holograph in which the pontiff's mouth seemed to open and close depending on the angle from which you viewed. Perhaps His Holiness was conferring a blessing, or merely yawning at the things he had to do now he was famous. All this was illuminated by a single light bulb suspended from the ceiling which, as Susan watched, tired of life, plunged to its death, exploded and assassinated the Pope.

A card in the corner of the window advertised: APOSTOLIC GROUP FAITH HEALING, and NURSE'S UNIFORM FOR SALE. Susan wondered whether a nurse's uniform might brighten our Sunday sing-song.

With eighteen mouths to feed, the weekly order of groceries was substantial. One of John's recipes began: 'Take the whites of 22 eggs . . .' So Susan made the usual cook's tour of the shops. Firstly, to Adams for ten pounds of ground coffee, then to the creamery for a couple of bottles of cream. There was a deposit on the bottles refundable on their return, but she didn't pay it. 'Ah there's no need,' the man said. 'If you're not to be bringing them back, just pop in later and pay the deposit.'

Fields the grocers – *'apples sold by the each'* – had a similar ethos.

Once Jack had asked for a small quantity of Arrowroot biscuits. They only had a large pack, so they split it open, Jack took what he wanted, returned the rest and received a refund. On the other hand, they boasted a marmalade so pure it contained no oranges, so John made our own at camp.

On one occasion the assistant asked: 'Would you like a fine salmon? Freshly caught,' he added with a wink.

'You mean it is poached?' said Jack, visibly shocked.

'No, raw,' the assistant replied with a smile.

The next stop was McCarthys for a seven-pound leg of pork and a shoulder of lamb. In the 1950s John had asked for four chickens and was offered a clutch of clucking feathers. 'They're still alive,' he observed.

'So they are,' the butcher agreed.

'I want to cook them, not keep them.'

With an exasperated sigh and a deft flourish, the butcher wrung their necks and plonked them twitching on the counter.

Fullers — *coal importers, manures, furniture, oil and ironmongery* — was famous (at least with us) for a unique type of toilet paper, unfoldable and non-absorbent, which in an emergency could be washed and re-used. Jack christened it 'Fullers' Universal' — the combined bum bumf and roofing felt. He found that, if crumpled, it worked as well as Brillo pads to make pans gleam or remove caked mud from shoes. The fussy, strait-laced little man behind the counter looked surprised at the quantity Jack was buying. 'Very useful stuff,' Jack confided. 'You would be surprised what I find to do with it.' The effect was electrifying.

We had ordered a new set of crockery, but they couldn't muster a complete set of twelve matching plates, cups and saucers, so they sent a mixture and were surprised when we complained.

'And two packs of candles,' Susan remembered before leaving the shop.

'Certainly, madam,' the deferential assistant replied. As soon as she left, he slipped across the road to purchase candles for her. For twenty-one years we had bought candles from a shop that didn't sell them, but was too polite to say.

Sadly, Fullers had got rid of their wonderful overhead railway for catapulting cash in hollow brass cannonballs across the shop from the counters to the lone cashier perched in her isolated office. The new front counter declared this was now a modern enterprise. But if you glanced over the assistant's shoulder, there in the back room was Mr Hegarty, wearing a dark suit and a high-wing collar. He sat at a tall Victorian desk and stooped over the accounts book like a vulture digesting its last meal. In response to an enquiry of Jack's he wrote: 'In advertance of mine of the 9th instant, we have found an interested party in regard to the type of prefabricated structure you desire.'

Apparently bereft of desire, Mr Hegarty had sat there for decades quietly adding up columns, subtracting and carrying three, his nib pecking at the paper. We had only ever seen him roosting at his desk in the back room. Perhaps he never moved. Then suddenly he did – he upped and opened a rival company.

On one of my rare visits into town I puzzled over the menu of enigmatic advertisements, such as 'Try Clarke's Plug, the Perfect Plug' (trial plugs?), or 'Wedding Ring Pillows Now in Stock'. Another stumper was 'Get Your Kosangas Here'. They had an African sound to them – ko-*zang*-ers – but I had no idea what they were. Fortunately, before I tried to buy some – 'Just one, please, in case I don't like them' – I realised it was, of course, Kosan *gas*.

I went past the Kosangas sign into Mr O'Sullivan's hardware

shop. After numerous complaints that it was impossible to use a microscope with a flickering candle, Jack bought a hurricane lamp. Overnight my microscopy was upgraded to that in the 19th century. Unfortunately, a careless splash of water shattered the glass. Mr O'Sullivan didn't have a replacement in stock. 'Take the glass from this one,' he insisted, and began to dismantle a lamp he had on display.

'Please don't,' I protested. 'With a bit missing you won't be able to sell it.'

'Don't you worry about that at all. I sold the wick last week.'

When I returned, a year later, less than half the lamp remained hanging from the beam.

I tried to buy a Gaz camping lamp. 'We're sold out entirely,' said Mr O'Sullivan, 'but I know a man up the street who has the very thing and he hardly uses it at all. Just you wait here and I'll be getting it for you.'

Skibbereen has always considered itself to be different, a Protestant island in an ocean of Catholicism. Sean O'Faolain called it 'the Wigan Pier of Ireland. County Cork taken to its illogical conclusion.' West Cork was justly considered a geographic narcotic. In the 1920s an IRA commander had claimed Skibbites were 'incapable of enthusiasm for anything more strenuous than promenading and gossiping', and a social survey in 1964 found 'a town lacking in initiative and enterprise'. Well, maybe, but the people were wonderfully warm and friendly. The price we pay for being go-ahead is too high.

It was unlike any town I'd known. When Jack's technician, Peter, once strolled through Skibbereen whistling, an old chap across the road took up the tune and began to dance.

Skibb has had many characters, but few famous sons. There

was Tim Sheehy, the 'Father of the Dáil' (the Irish Parliament), and Humphrey O'Sullivan, who went to Boston and invented rubber soles for shoes so that nurses could creep up on their patients. In 1899 he founded The O'Sullivan Rubber Heel Company and became rich.

The happy anarchy of the South fermented rebellion against Britain, and O'Donovan Rossa, although living in the most Protestant town in the south, gave birth to the Fenian movement. His death was one of the sparks that ignited the Easter uprising of 1916, because of Patrick Pearse's oration over his grave:

'The defenders of the realm . . . think they have pacified Ireland . . . but the fools, the fools, the fools – they left us our Fenian dead, and while Ireland holds these graves, Ireland unfree will never be at peace.'

Although 49,000 Irish fighting men died in the Great War, it was when the war ended that the conflict in Ireland began. With Irish understatement, the bloody civil war was called 'the troubles', as if it were just a tiff between neighbours. The King's Liverpool Regiment and a detachment of the Black and Tans were billeted in O'Donovan's shop in Market Street. They occupied the upper storey while, unknown to them, Sinn Féin stored their most sensitive documents on the floor below.

Skibbereen fired off more words than bullets, so the police shut down the nationalist Catholic *Southern Star* for 'inciting disloyalty and disaffection to his Majesty'. Then the loyalist Protestant *Skibbereen Eagle* had its presses flung into the river by 'armed and disguised men'.

By 1920 the whole of the south-west was under martial law. A British sailor patrolling the coast spied 'nothing but rocks, sea and Sinn Féiners'. Over half of the British army in Ireland was occupied in County Cork. It struggled to subdue seven battalions of the IRA. The Fourth Battalion was organised around Skibbereen, although only four townsmen were members. Tom Barry, the local guerrilla leader, described Skibbites as 'a race apart from the sturdy people of Cork . . . slouching through life meek and tame, prepared to accept ruling and domination . . . providing they were allowed to vegetate in peace . . . If Satan himself appeared in Skibbereen . . . the majority would doff their hats to him, and if he wagged his tail once in anger, he was sure to be elected . . . to the District Council.' The *Skibbereen Eagle* saw the Volunteers on parade as 'Skibbereen sparks wishful of a change from billiards'.

Guerrilla war raged in nearby townlands, with the IRA following Jonathan Swift's instruction to 'burn everything English except their coal'. But Skibbereen remained relatively quiet because, while torture and summary executions by the British were commonplace elsewhere, the commander of the Skibbereen garrison imposed a strict code of conduct. He even rescued prisoners from the bloody hands of the Black and Tans. The rebels responded in kind. They made occasional forays into the town to fire on army posts and burn down the courthouse, but took captured soldiers off to a pub, plied them with whiskey, then sent them home singing rebel songs:

> We've got the guns and ammunition,
> We know how to use them well,
> And when we meet the Saxons,
> We'll drive them all to hell.

With the truce of 1921 the army withdrew and the real trouble began. The Treaty of Partition was signed and civil war broke out between the forces of the newly-formed Irish Free State commanded by Michael Collins ('the Big Fellow') and republican rebels led by Eamon de Valéra ('a cross between a corpse and a cormorant'). Skibbereen barracks were taken over by Government troops, so Republican 'Irregulars' mounted an unreliable cannon on the top floor of the Bank of Ireland across the street. Although it battered the barracks, it also blew the roof off the bank.

Skibbereen had always been a place to leave. Almost every venture failed and even the railway closed down. 'It's a fine place to commit suicide in,' a local boasted. There had been one setback after another, but one disaster was the worst of all . . .

On 1 July 1845 Father Mathew left Cork City and travelled through fields of potatoes that 'bloomed in all the luxuriance of an abundant harvest'. When he returned only seven days later, he 'beheld with sorrow one wide waste of putrefying vegetation'. All over Ireland the plants perished, every potato turned to slime. It was the deadly murrain . . . vegetable gangrene . . . potato blight.

Year after year the crops rotted in the fields. There was no cure and no salvation. 'We are visited by a great calamity which we must bear.'

Eye-witness descriptions of the suffering around Skibbereen are beyond belief:

'Famished and ghastly skeletons . . . two hundred such phantoms, such frightful spectres as no words can describe.'

'. . . a rampart of human bones . . . this horrible den . . . a mass of human putrefaction.'

'Seven wretches . . . under the same cloak, one had been dead many hours, but the others were unable to move either themselves or the corpse.'

'Dead children lying by the roadside . . . nothing can exceed the deplorable state of this place.'

The sole employment was on public works. The Rapids Quay at Lough Ine was repaired, but much of the work was designed to be as repulsive as possible to discourage malingerers.

Some worked and others just lay in the streets, but either way they died. By the time the famine was over, the population of the district had been halved and, in Ireland as a whole, one and a half million had perished and over a quarter of the population had emigrated. It became the greatest migration of people in the history of Europe. It was 1966 before Skibbereen arrested the decline in its population. Even today Ireland has only two thirds as many people as before the famine. Strangely, even the Bohanes, who had tales about everything, had none of the famine.

Some of the survivors of the 'coffin' ships to North America carried a sprig of hawthorn or meadowsweet from home to put down roots in the New World, and thus Irish America was born, with a legacy of bitterness to nurture.

Oh son! I loved my native land with energy and pride,
Till a blight came o'er my crops – my sheep and cattle
    died.
My rent and taxes were too high, I could not them
    redeem,
And that's the cruel reason I left old Skibbereen.

Oh, father dear, the day may come, when in answer to
    the call,
Each Irishman, with feeling stern, will rally one and all.
I'll be the man to lead the van beneath the flag so
    green,
When loud and high we'll raise the cry – 'Remember
    Skibbereen'.

Beyond the town, on the banks of the sullen river Ilen, are the ivy-laced ruins of an old Cistercian abbey. A wrought-iron lattice in the grounds marks a mass burial, which was described by O'Rourke, the Irish chronicler:

'Monster graves, called by the people "the pits", into which bodies were thrown coffinless in hundreds, without mourning or ceremony – hurried away by stealth, frequently at the dead of night, to elude observation and to enable the survivors to attend the public works next day, and thus prolong for a while their unequal contest with an all-conquering famine.'

The pits were kept open until full, flaunting hundreds of bodies shrouded only in a scatter of sawdust, their limbs deliberately snapped for closer packing.

When he first came to Lough Ine, Louis Renouf met a man in his eighties who was crippled by a childhood accident. Thought to be dead, he had been flung into the pit. The jolt of having his legs broken with a spade revived him, and he was hauled out over the immense mound of corpses.

Predators

Through the long hungry years, few realised that they lived beside a larder; adjacent to the barren land lay the fertile but little-used sea.

As the tide recedes, it uncovers catchable parcels of protein called crabs and prawns, mussels and winkles, but most Irish peasants had no idea how to prepare or cook them. They remained unharvested, at least by humans.

Marine predators never fail to recognise food, no matter how it's packaged, whether finned or footed, shelled or slimy. The female herring lays a million eggs to ensure that at least two survive to maturity and keep the species going. Almost every organism is destined to be devoured. Such deaths are the currency of life.

At first, Jack and John believed that the distribution of marine creatures was explicable in terms of factors such as tides and waves, but gradually they changed their minds, and it was this that was to make their work famous.

They began to realise that interactions between the organisms themselves were at the heart of what was happening, and that certain key predators or competitors exercised an over-whelmingly important influence over the entire community. Their research switched to a search for such key organisms.

Long before I came to the lough they noticed that the common mussel was abundant only on the most surf-battered shores and in the calmest corners of the lough. Surely it could inhabit many of the sites in between — so why wasn't it there? To find out, they transplanted mussels to intermediate shores and, within two days, they had been consumed by marauding crabs and starfish. For the first time, the team fully appreciated the difference between what distribution was possible and what was permitted by other creatures.

The sea slithers with fearsome predators. Once a mussel is embraced by a starfish, it is lost. Starfish use hydraulics instead of muscles, and in a tug of war they never tire. They can drag open any shell and, as soon as it gapes, the starfish throws out its stomach like a white shroud and dissolves the victim alive in acid. If a starfish wanders into a pool, the resident urchins soon 'smell' it and stampede to the other end, for they know what is in store. I rest my hand in the pool and the starfish folds itself around me. It senses I am meat.

The shore is the home of quiet murders. Sea anemones only masquerade as harmless flowers. Each tentacle is laden with ten thousand hypodermics loaded with poison waiting to paralyse the unwary. The tentacles of the snakelocks anemone stick to my finger, held by its barbs, but unable to breach the skin. There is safety in size. If we were small, the shore would be a terrifying place and our first seaside holiday would be our last.

Several projects confirmed the importance of predators. On the exposed coast at Carrigathorna, for example, the shells of the dog whelk were thin-walled and wide-'mouthed', whereas in sheltered sites they were sturdy with small 'mouths'. A 'nut-cracker' device showed that the sheltered shore types were more than twice as difficult to smash. This seemed to be the wrong way round, surely they needed to be tougher where the surf would pound them. Not so, for even the more delicate form could withstand an occasional tumble around the rocks when dislodged by waves, and the large aperture of the shell allowed it to have a larger 'foot' to fix securely to the substratum. When the small-mouthed forms were transplanted to Carrigathorna most were lost within a week. But why were the sheltered water inhabitants so heavily armoured? It was because crabs, which shun surf-washed shores because they would be washed away, grew big in calmer sites and, although easily capable of smashing the flimsier-shelled whelks, they had a hard time with the tougher ones. Even so, the whelks sometimes retreated up the shore to risk death from desiccation or rain rather than be eaten alive. Again, predators were calling the shots.

The work at Lough Ine became a hunt for other examples of biological control. By manipulating nature – adding or subtracting predators or competitors and following the repercussions – they were able to identify those key species that controlled the success or failure of the others. Organisms were moved to areas of greater or lesser risk. The number of predators or grazers was adjusted by hand or excluded with cages (although at night, mischievous otters sometimes unscrewed the lids and stole the contents). Jack and John discovered not just who was eating what, but also how big a predator needed to be to devour prey of various sizes. Small juvenile animals were more vulnerable, but might evade capture beneath the adults or in the hidden corners of the habitat. Most important of all, they revealed that a cascade of consequences followed from changes in the abundance of a single species.

Long ago it had been observed that if the leg of a dead frog was stimulated with an electric current, the muscle contracted, and this provided a way to study how muscles work. What Jack and John developed was a means to goad living communities into revealing how they operated. Their pioneering approach showed that ecological experiments could be carried out in the sea. An entirely new natural laboratory had been discovered, one that avoided the artificiality of laboratory-bound experiments, which were the mainstay of the rest of biology. Their first publications on the predation of mussels (in 1959) and the importance of urchin grazing (in 1961) had ushered marine ecology into the experimental age and their techniques allowed ecologists worldwide to use shore-dwelling organisms as test beds for ecological theories of how communities developed and were maintained.

Identifying the most influencial species was also vital if we were to predict what would happen to coastal communities should the oceans warm, or an oil spill kill off the limpets, or a new sewage outfall coat the rocks with silt.

Thanks to my studies at Lough Ine, I too now embraced the significance of biological interactions between organisms and learned how to unravel them. I became a workman who had found his tools, and my PhD thesis would reflect this change in approach; the first half was about the distribution patterns of *Saccorhiza* in relation to factors such as temperature, light and water flow, but I decided that for the remainder I would examine the most important creatures that lived on or around the plant and how they could overwhelm or exclude it. I had begun my explorations on the effects of competition, and how grazers select their food, something that would last me a lifetime.

The most important grazer in the lough was the purple sea urchin. This is a Mediterranean species that reaches its northern limit in Ireland. On the soft limestone shores of the west coast it lives in flask-shaped depressions in the rock. It crawls in when young and gradually enlarges the 'flask' without widening the mouth, so that pretty soon it is trapped and, unable to browse, has to rely on scraps of seaweed drifting in and getting caught on its spines. Here in the lough it just crawled around between and on top of the rocks, chomping the algae. It was conspicuous in the shallows during the day, but hid beneath the rocks before its enemy, the spiny starfish, came out to play at night.

The urchins controlled the distribution of their food and therefore determined the character of the shallow underwater communities in the lough. In places the rocks were bereft of seaweed, where herbivorous urchins were abundant – but when Jack and John removed the urchins a lush vegetation of soft, succulent algae arose. When the animals were replaced they formed themselves into groups, eating their way outwards in expanding circles, leaving bare rock behind. It was the first evidence that urchins were the 'cows' of the sea floor. Now we know of many places where there are 'urchin barrens' underwater, because no seaweeds can survive the grazing pressure.

I was puzzled why *Saccorhiza*, which was so abundant in the Rapids, was absent from inside the lough itself. Were grazing urchins responsible? I set out to test whether they would eat *Saccorhiza* and offered pieces of the seaweed to urchins in submerged cages. I was dismayed to find they didn't eat the algal swatches, but wore them as hats instead. So I supplied them with oyster shell fedoras, and then they tucked into the plants. Were they really more fashion conscious than hungry? I offered the urchins a choice of two different seaweeds and they invariably scoffed *Saccorhiza* in preference to the other, a kelp that *did* occur naturally in the lough. I tasted them too, but preferred the other one — so the urchins were clearly discriminating, but not discerning.

The urchins living in the lough rarely left home, especially in the afternoon, without donning a shell, a leaf or, for special occasions, a tuft of algae. In the evening they rarely bothered. No one knows whether these decorations are parasols against the sun's glare in the shallows or camouflage to disrupt their outline as seen from above by potential predators such as birds. Hats are, of course, no disguise against their main enemy, crabs, but if there were no benefits to be had, why would they waste time fiddling with millinery instead of gliding off bare-spiked to browse and build up their gonads, to ensure that there would be more urchins next year?

Urchins had long been one of the most abundant creatures in the lough. If you stepped in the shallows, you were lucky not to get skewered in the foot. The brittle points break off in the skin and the wound festers if you don't get them out. I suffered for days after being spiked, until I realised that spines still lingered in the rubber bootees of my diving suit. Only when I put them on and my weight compressed the soles did a fakir's mattress emerge.

We had taken the urchins for granted, but this summer they were almost gone. Our census showed that the number in the southern basin of the lough had fallen from 35,000 a few years before to just 3000. In their absence, *Saccorhiza* had not invaded the lough, but Bunny's seaweed, *Codium*, had flourished and now painted a bright green swathe throughout the shallows – except, inexplicably, in what had long been called Codium Bay. I showed that the urchins ate the weed and destroyed far more than they consumed, by nibbling through the base so that the whole plant was detached and lost. A sea slug also eats *Codium*, but absorbs rather than digests its packets of chlorophyll, moves them to just under its skin and becomes a photosynthetic animal; it utilises the 'food' that continues to be produced by the captive chlorophyll.

But what had caused the urchin decline? The urchin populations in Bantry Bay, just along the coast, had gone to French dinner tables, but there was no suggestion that they had been harvested from Lough Ine. Jack suspected crabs might be involved, so we set crab traps. When lecturing on this work a few years later, I accidentally called them trab craps, and found that once reversed, the words refuse to go back. As a result, the lecture went exceptionally well.

We baited the traps with Kit-e-Kat cat food and, after opening a dozen cans, the students turned decidedly fishy. The aroma drifts down the years; I can smell it still.

The number of crabs had increased tenfold over the last

couple of years. Probably the mild winters had allowed them to stay in the shallows eating urchins instead of retiring into the depths as they usually did. But could even the aggressive devil crab destroy urchins on a sufficient scale? To find out, we moved a thousand urchins into an urchin-free area beside the Glannafeen laboratory. That night I dived over the site and found a battlefield with dead urchins everywhere. Crabs are assassins in pie crusts. The big ones can smash an urchin with a single blow, but the little ones have to juggle, turning the urchin over so they could insert their claw into the soft tissue around its mouth, then twisting it round and can-opening it. These deeds are so dirty that they are only perpetrated under cover of darkness. In a single night, eighty per cent of the transplants had perished. But why do crabs bother with urchins, which contain less edible tissue than almost any other creature?

Because little is known of the natural fluctuations that occur from year to year, it is difficult to determine whether changes that follow a pollution incident, for example, result from the pollution or would have taken place anyway. We needed to measure the range of nature's normality as a yardstick for comparison with the effects of other perturbations. Jack decided to monitor the abundance of urchins and *Codium* every year from now on to see if a severe winter reversed the trend.

I helped out by supervising the *Codium* census. The assessments were no fun. Before the plants could be weighed they had to be spun in a net above your head twenty times to remove excess water like raindrops from a twirling umbrella, and by the end of the day some students couldn't lift their arm above shoulder height. They got their own back by serving me a *Codium* and cream pie, which I ate, but never ordered again.

Jack also got his come-uppance at the Sunday sing song:

> If you go down to the lough today you're sure of a
>   big surprise.
> The purple urchins have gone away and *Codium's*
>   thick as flies.
> And all the students that we can spare,
> Are weighing plants and tearing their hair,
> The students will wear, but Jack doesn't care . . .
> He's having a picnic.

The way we tackled this project was a perfect example of Jack's ingenuity. We needed to estimate the total amount of *Codium* in the entire southern basin of the lough, a seemingly impossible task. Fortunately, the *Codium* zone was a narrow ribbon in shallow water, following the contour of the shore. So all that was required was for a boat to be rowed along the length of the zone with an observer hanging over the stern holding a two-metre rule, to measure the width of the zone at regular intervals and estimate the percentage of the sea floor within the zone that was covered with *Codium*. Then we weighed all the *Codium* plants in sample areas with a wide range of *Codium* cover to convert cover into weight and calculate the total weight of the crop in the south basin.

Nowadays we would employ aerial photography and interpret the pictures with computerised image analysis and spend a fortune in the process. Jack's genius was to do the job just as well using a rowing boat, a long ruler, a net bag and a spring balance. The total cost of all the equipment, including the boat, was less than £50.

To facilitate the logistics of estimating seaweed cover and counting urchins, we needed to divide up the shore into manageable lengths.

'We should re-mark the Renoufian sectors,' said Jack.

'Good idea,' we all agreed. 'What are the Renoufian sectors?'

'Ah!' said Jack. 'In the 1920s Professor Renouf surveyed the shores of the lough and described eighty-seven different stretches of shore. He numbered every outcrop, beach and bay. If only they were clearly marked it would facilitate our surveys, don't you think?'

So I set off with Norman, an unsuspecting student, and a tin of yellow paint. I rowed the boat with one oar permanently on the shore – a little used art – and re-described the sectors:

South 6 – Rock wall to Mermaid rock.

West 18 – Promontory to thin upright rock with 'bullet' hole.

West 37 – Tombstone to boot-shaped promontory.

Jack always named the sites where we worked, and over the years almost every inlet and promontory was christened: Eddy Creek, Graveyard, Rookery Nook . . .

At intervals I deposited Norman ashore to paint small crosses on the rock to delimit each sector. Stepping back to admire West 37, he slid gracefully backwards on the slick seaweed, then flipped over – with pike and half twist – and emptied an entire can of daffodil gloss over his head. He looked just like someone who had met an alien and been turned into a yellow putrescent mass.

Jack's first-aid advice was usually limited to 'Pack it with Savlon', but this time he bellowed, 'Turps him all over, then hose him down.' We did, and his skin came clean, but not his buttercup hair. In an instant the punk movement was born, although it took fifteen years to catch on in London.

We wondered what tourists might make of the rash of small

crosses around the lough. Would they assume they marked the spots where a surfeit of saints had flung themselves into the purifying waters in a mass baptismal rite?

Perhaps they did indeed have a magical quality, for thirty years later they would still be clearly visible.

Wildlife

We had other crosses to bear. The students were mostly city folk and, except for the toddler's tepee in the garden, had never lived in the wild. Fortunately, one of the them was a Queen's scout. He knew all about tents and was forever fine-tuning the tension of the canvas by tightening and slackening the guy ropes in response to the weather. But one day he tightened when he should have slackened. At midnight, while he and his three companions slept, the guy ropes shrank and the canvas became as taut as a drum skin. Suddenly it ripped away from the apical collar that held it on the pole, and the heavy tent slumped on to the terrified occupants. It took over ten minutes to free them.

Being properly equipped makes a big difference when camping, and between them Jack and John saw to everything. Jack was very precise about what clothing was required by the happy camper. Students could be relied upon to bring daft hats and high heels or a bathing costume and sun oil, but Jack ensured they were ready for the realities of rural life:

Things needed

1. Warm clothing and still more in case the first lot gets wet.
2. Waterproof hat with brim, e.g. sou'wester, to stop the rain from going down your neck.
3. Waterproof clothing, e.g. oilskin. It should reach to thigh length. A jacket is not enough unless combined with waterproof trousers.
4. Plastic mac – useful for walking, but the buttons rip off in a boat.
5. Rubber boots – needed around camp in wet weather and on the shore.
6. Old gym shoes suitable for paddling in as protection from the spines of sea urchins which break off in your foot.

It is difficult to be green when in greenery, but fear not, we had rituals. Fourteen people produce a lot of rubbish and the disposal procedures were laid out along military lines. Coffee grounds and washing-up water were flung into the bushes, along with the inevitable teaspoons. Fortunately, John collected 108 coupons from packets of Erin peas and exchanged them for three dozen new spoons.

We put biodegradable kitchen refuse into brown paper sacks so that it could be dumped at sea. It was strictly forbidden to put wet coffee grounds into the sacks as they caused the bottom to fall out and fill the boat with refuse – but everyone forgot. We might just as well have slung the rubbish directly into the boats and saved on sacks.

Objects that didn't rot had to be returned to Skibbereen. The butt end of a pickaxe handle was used to flatten tins, while simultaneously squashing toes.

In the wild, the convenience may be inconvenient. On the hill above the camp stood the latrine, an open trench over sixty centimetres deep, surrounded by a hessian screen that was held up by onion-topped poles. Near at hand was an old galvanised watering can full of creosote. Although the pit was watered on every visit, nothing ever grew except unease. Fresh from *Quatermass and the Pit* on television, I suspected that some malevolent 'thing' lurked below. It was the look of it, like seething magma craftily quiescent whenever you glanced. But there was always the danger that, while we slept, it might emerge and slither downhill to engulf us.

In the heat, the main danger was from biting insects, attracted when you were at your most vulnerable, hobbled and astride the trench. The Queen's scout had the answer; wildlife flees from fire, so a burning brand of rolled newspaper held above the head should do the trick. It worked – and then it worked even better, for he dropped the fading torch into the pit, the creosote ignited and WURRUMPH! Mount Latrine erupted. He screamed and leaped clear as the creosote-soaked hessian screen caught fire and was consumed in seconds. The scorch-blackened poles hesitated for a moment, then one by one fell into the trench. It was a cartoon catastrophe. Our Queen's scout was taken aback, depilated, but undeterred. Next day, bandaged as befits an Egyptian mummy, he was shuffling between the tents tightening and slackening the guy ropes.

One morning we were awakened by cries of 'Help! Help me!' The replacement screen for the latrine had begun to collapse in the wind. A student in residence at the time clung desperately to the tilting poles.

It was only a short, steep step up the hill when it was dry, but when it was wet it was horrid. The next year it was wet.

For our Sunday sing-song I penned a new version of the 'Chim-chimeney' song from *Mary Poppins*:

> Each year at Lough Ine I love the work that we do,
> But the reason I come is the view from the loo.
> The students complain that they can't get through,
> Some slip away and we lose quite a few.
> Slip slippity, slip slippity, surely it's true,
> The treat you can't beat is the seat at the loo.

Later we installed a chemical lavatory, a bucket of blue Elsan disinfectant, and I mourned the passing of the little trench upon the hill:

> In the cool of the evening my memories burn,
> For the hessian heaven that will never return.
> Now covered in Elsan, I'm feeling blue.
> It just can't compare with the loo with a view.

The new chemical latrines had none of the sensory luxuries of a real lavatory. Felicity, one of the students, expressed this longing when she said with a sigh, 'Oh for the music of a flush.' So I hung an impressive but inoperative toilet chain from the rafters and almost everyone automatically gave it a yank, then swore at the dope who had put it there . . . and the dupe who had pulled it.

Camping brings you closer to nature. Sometimes too darned

close. Safari beds suspend you just fifteen centimetres above the soil, and it's not enough. Even the city slickers who could have Hush-Puppied their way through the meanest streets were unaware that the wilderness writhes with all manner of creatures, almost every one of them repellent. Each beast and bug is slime-secreting or laden with toxins or has an urge to scuttle up your trousers.

Nor were we all brave. One girl looked down a microscope and screamed because something moved.

Country spiders were twice the size of those that lurked in corners back home. Slick liquorice slugs slithered by, unravelling slime behind them. In bisexual abandon, they hoved alongside like men o' war and fired fusillades of sperm shells into each other's genital pore. It was like the Battle of Trafalgar. There were grey slugs too, eight inches long, surely an aesthetic abberation. When courting, they climbed a tree and circled for a couple of hours eating each other's slime as a token of affection. Then they wove a long cord and mated while dangling in mid-air. 'Oh dear,' said a local when I told him. 'If only the Lord had rested a day earlier, we would have been spared all this.'

The area was rich in wildlife, unafraid of man – one day a robin hopped on to my shoulder to see what I was reading outside my tent. This really was the *great* outdoors. Larger beasts were no bother, although rats unsettled our sleep, otters stole our experiments, and then there was the shark . . . To rattle strangers, Alan, one of the students, had attached a triangular 'fin' to a hydrofoil. When Jack rowed visitors back to their car at dusk, the towed fin diving and snaking behind the boat was frighteningly convincing. 'Merely a shark,' said Jack, casually. 'Often follows us – quite safe in the boat.'

*

The nearest the students came to being naturalists was as fisher-men. Although the largest mackerel ever caught in Ireland was hauled out of Lough Ine, it was certainly not caught by students. They usually failed to catch anything even when fish were leaping from the water all around, trying to give themselves up. If by mistake something *was* hooked, the anglers were excused washing up.

A student with an evangelist's zeal (a fisher of men) was also unsuccessful, although he was a believer in miracles – he gave Jack a Bible. Jack accepted it graciously, but I doubt he ever consulted it. The student with ornithological leanings, however, was the worst kind of untalented enthusiast. The idle were no problem, they were merely useless, but energy allied to inability caused all sorts of trouble. Fortunately his bird-watching gave him away. He was forever spotting rarities; among a squadron of sparrows he would spy a fulvous babbler (which turned out to be a thrush), a Tennessee warbler (a garden warbler), or a chough (surprisingly, a chough).

At dusk he stationed himself beside the tent door with flaps opened wide, so that his ever-alert binoculars might observe evening grosbeaks without fear of correction. One night he fell asleep slumped across the door and the camp was awakened by screams. A cow had wandered into the field and mistaken his moonlit face for a salt lick.

There was more wildlife at Lough Ine than Jack imagined. It is one of the miracles of life that each generation of adolescents discovers something wonderful that is quite unknown to their parents. I know of one or two students who misplaced their virginity at the loughside. One of them went on to write a book called *Sperm Wars*, which surprisingly made no mention of Lough Ine. Even the students' official diary, that would be typed-up by

Jack's secretary back in Norwich, contained a reference to pro-
longed sunshine that resulted in 'sunburn to the back of Keith's
legs and Claire's knees'. As close proximity and limited
opportunity conspired, the pressure mounted. It resulted in at
least one instantaneous female orgasm and I was lucky enough to
be present at the time. Jack, the guardian of lost morals, would
have been horrified to learn that anything so un-Enid Blyton-ish
as *Five Go Fondling* had occurred; John would have been surprised
it hadn't happened more often.

However, if you think that lusty young students were forever
sneaking off into the undergrowth to practise rolling hitches, you
have reckoned without the dread of sheep ticks.

Ticks are tiny eight-legged black discs, at least that is how
they start out. When they have gorged on blood, their abdomen
inflates until it is a grotesque pale sphere. In the absence of
sufficient sheep, cows or rabbits they are happy to harpoon
humans. They climb to the tops of the vegetation . . . and wait.
Special heat sensors locate a passer-by and the tick grabs hold.
Before it digs in, it sniffs to ensure you are a suitable host – but
don't worry, you are. Once on your legs, it ascends. Whether they
like warmth or darkness I don't know, but they don't stop until
they reach your pubes and dig in their large mandibles.

There can be as many as two and a half million to the square
mile. The female sucks up over a hundred times her weight of
blood before dropping off, sated but content, to lay 10,000 eggs.
These hatch into tiny larvae all eager to climb up into your shorts,
for they too must taste blood before they can move on to the next
phase of their life history.

Tick 'teeth' are twisted to the right, so if you pull the beast
out straight, the jaws stay behind and fester in the skin. A sharp
anti-clockwise twist is preferable and it is best to relax the

creature first by singeing its bum with a cigarette or immersing it in paraffin or alcohol. Alcohol enthusiasts can become so relaxed that they fall over – ticks fall out.

Embedded ticks don't irritate their host, and so go undetected. I recall a student absentmindedly scratching inside his shorts and then, like a conjurer, producing a bloated tick as big as a chick pea. To avoid such embarrassments, every night there was a tick inspection of legs (etc.) with a torch in one hand and a tube of meths in the other. Renouf used to offer his children a penny for every ten ticks they collected. We offered a bottle of Guinness to the camper who found the most. It was a male preserve, for the hairiest legs were the best collectors. The record was thirty-seven in one day and, although we suspected he had rolled in the bracken to overtake the favourite, nobody begrudged him the prize. I once knew a researcher who worked on body lice and kept them strapped to his wrist in a little bottomless box in case they felt hungry. It takes all sorts.

Some horseflies are three centimetres long and have huge jaws, so when they bite, you know you've been bitten. They can even irritate thick-skinned cows, who on dry days perch on the tops of hills where the wind blows the flies away. But Glannafeen was sheltered from the breeze and we were being eaten alive. Bunny reacted especially violently, and once returned home decorated with angry red welts.

It is only the female that sucks blood (in mosquitoes too it's the lady who bites). The gentle male sips nectar while she, with scissor jaws, slices into her victim's skin and then prises open the wound like a practised surgeon. It takes only a second and is instantaneously excruciating.

Although unable to tell a marine biologist from a cow, she

has no difficulty distinguishing a knee from an elbow. The three commonest species of horsefly divide up our bodies between them: the zebra-striped one goes for the face and neck, and the really big one takes the legs. This leaves your arms and back to the drab, but eager, cleg. The cleg was the worst; once it fixed you with its iridescent eye, no amount of windmill waving of the arms would put it off. Its only endearing quality is that it becomes so engrossed in its dinner you always get it, and revenge is sweet.

It must be great to be an entomologist – no wading, no diving, no searching, just strip off and lie there and the samples fall into your lap. Well, hopefully, not your *lap*.

With such a selection of horrors to banish from Ireland, why did Saint Patrick pick on the poor frogs and snakes?

'The horseflies are terrible,' I said to Jack.

'Elephant flies are worse.'

I crooked a doubting eyebrow.

'You haven't seen the elephant fly?' said Jack in his usual deadpan voice.

'No,' I admitted, 'but I once tripped over its droppings.'

Sparks

When I had first arrived at the lough only a couple of years before, the locals had spent much of the day going to the creamery as they had always done. The lanes were busy with donkey carts each carrying a single churn of milk. 'We spent hours waiting in line,' John Bohane told me. 'The separator was always breaking down. But it was a fine chance to gossip and tell tall tales. If you fell in with the right fellow, you'd have a great day o' wit.'

By 1966 the donkeys were in decline. Most local farmers were supported by government grants and drafts from relatives in the States. 'Sure 'tis lucky for the Bohanes that Columbus discovered America,' John Bohane admitted, although I suspect he was just kidding.

In the last year, Jack had transformed their lives. He had 'the electricity' brought in for the Dromadoon laboratory and a solitary cable looped its way across the hills on a line of poles.

The Bohanes tapped into it and their parlour was now transformed into an electricity showroom with a washer, a refrigerator, an iron and electric light. Philomena had discarded home-baked soda bread for a freezer full of white sliced.

It seemed perverse that Jack had electrified the entire neighbourhood and brought power to a laboratory we used infrequently, while our main lab at Glannafeen and the mess hut where we cooked, ate and wrote in the evenings remained lit by candles. Perhaps Jack loved the soft trembling light they gave, or maybe our special little world would have faded in the glare from an electric light bulb. When I raised the matter ever so gently, he just said, 'No, not necessary.' So I never raised it again.

John Ebling left half-way through the expedition to fly to the United States for a few days to give a lecture.

'He often disappears abroad during our summer expeditions,' Jack explained. 'On visits to the Tiger Balm Gardens, skinbinges in the USA, hair-affairs in Mexico or high-level conferences at Delhi airport.'

Last year John had claimed that 'Delhi airport is so busy these days I invariably meet someone I haven't seen for years. I reckon if I were to linger there for a couple of hours I'd be certain to encounter a colleague. Uhum!'

'Really?' Jack had said, with doubt dripping from his lips.

Six months later he passed through Delhi and sent a cable to John from the airport. It merely said: 'Where on earth were you?'

They both liked to travel. Jack took his wife abroad every winter, perhaps to make amends for hiding away at Lough Ine each summer, whereas John travelled so much because pharmaceutical and cosmetics companies were happy to despatch him to seminars and conferences around the world. I think Jack disapproved of such 'disreputable' sponsorship.

John was missed, for thanks to his recipes we had already dined on lobster, a huge leg of pork encrusted with crackling, and salmon and clams cooked in sherry. Our cook coped well without him, but Jack could not provide his culinary flair. Apart from Guinness omelettes, his speciality was spotted dick with most of the spots omitted.

A large joint was delivered unlabelled.

'It's hogget,' Jack declared confidently.

So we treated it as lamb for roasting; unfortunately it was ham for boiling.

'Glad to see you safely back,' said Jack when John returned. 'No doubt you will have added to your already over-long list of improbable stories.'

'Just a couple of amusing misadventures,' John admitted, modestly. 'When I was at Niagara Falls, would you believe—'

'Probably not,' Jack said.

Over dinner I updated John on what we had been up to while he was away. 'We've harnessed the new electricity supply at the Dromadoon lab and installed a pump to lift sea water from the Rapids and flush it across a waterproof bench so that animals can be kept alive.'

He thought for a second then said, 'Have you passed water yet?' I laughed and John couldn't resist turning to Jack, raising an eyebrow, and saying, 'Uhum!'

It was a mistake. Jack turned puce and leapt to his feet. 'I want to speak to both of you next door!'

The headmaster marched the two naughty schoolboys (John was forty-eight years old) into the laboratory and gave us a wigging. 'I will not have that sort of suggestive chat in front of these . . . these tender plants.'

I assumed he meant the students, but it was not my assessment of them.

Such an outburst over an apparently minor matter would have only astonished and amused me, had it not been for the way in which Jack's relationship with John, a collaborator for over twenty-eight years, had instantly resorted to its original status – that of master and student. But if he was upset, John didn't show it. He merely smiled at me and shook his head almost imperceptibly. Nonetheless, I realised for the first time that even apparently calm waters may hide submerged mines.

The next day we were working in the shallows lifting rocks into a boat. One slipped from my hands and splashed a female student.

'Shit!' she cried. 'I'm f***ing well soaked.' She then removed her T-shirt to wring it out.

Jack looked up and said nothing, but must have realised that the students were far hardier than he thought. For all his friendly informality with the students, deep down he was an Edwardian headmaster in charge of Saint Trinian's, and there was nothing he could do about it. Although I didn't realise it at the time, the gulf between him and the students could only widen.

But at least he never called them tender plants again.

✿

In 1966 I had to leave the camp early to rush back to Liverpool
and finish my thesis. But I was late arriving in Dublin, so stupidly,
I told the taxi driver to hurry in case I missed the ferry. It was as
if he had Vaselined the sides of the cab so he could slide past the
startled traffic. Suddenly we swung off the main road and swerved
to avoid a barrier: NO ENTRY – ROAD UP.

We rocked and rattled over a surface of rubble and
brickbats. The taxi was shaking apart.

'Don't worry,' said the driver. 'It's only a short cut.'

'No rush,' I said, but to no avail.

I again found myself in front of the Customs and Excise
man. It may even have been the *same* man as last year.

'Been in contact with foot and mouth?' the officer asked
sternly.

'No feet, but one or two mouths,' I retorted smartly.

He dismantled my luggage and, as a result, I was the very last
passenger to board the ship.

The writing-up went fine. By the end of the summer I had only
one more chapter to go when Bunny announced, 'I am going on
holiday to my cottage in Dorset and taking Miss Price with me.'

'What about my thesis?' I asked pathetically.

After considerable thought she said, 'I suppose you can
come too.'

In Dorset, Bunny and Win were out all day collecting algae
while I laboured over chapter six. In the evenings Bunny cooked
excellent dinners and filled the house with the aroma of warm
loaves. 'I love making bread,' she admitted. 'It gets my hands so
clean.'

The ladies slept in the delightful thatched cottage, but the rogue male was relegated to the dank garage outside. The roof was half intact, and only after I almost drowned in a nocturnal thunderstorm was I allowed into the house.

In those days, the universities were expanding and jobs were plentiful. I secured a post as a junior lecturer at Glasgow University even before my thesis had been examined. Two months into the job I had to return to Liverpool for the viva. It went well, and I didn't need to stay over for a second night as I had planned, so I took the evening train back to Glasgow.

The departmental secretary who changed my booking gave my name as *Doctor* Norton. In the middle of the night I was awakened by the attendant. 'You are a doctor ain't you, sir?'

'Well, yes,' I replied modestly. He must have heard it announced over the radio. 'In philosophy,' I was about to add, but it was too late.

'Come with me. It's an emergency.'

I scuttled after him, still trying to explain.

'This lady is having a baby,' he said, gesturing to a writhing figure lying on the bunk. 'I'll leave her in your hands.'

I whispered the terrible truth into his ear.

'Never mind sir, you're better than nothin'. Anything I can do?'

'Look for a fat lady with ten children, even a slim one will do. And boil some water.' It came instinctively. In Westerns the grouchy old doc always took one look at the woman, rolled up his sleeves and shouted, 'Hot water and plenty of it!' Presumably she had to drink it to replace lost fluids.

'Yes, sir. Right away, doctor,' he replied and obediently rushed off leaving me with the moaning woman in mid-contraction.

What on earth was I to do? I had never even seen a cat giving birth to kittens and I get giddy at the slightest sight of blood. What if she asked me if I had a speciality? How would she take it when I said seaweeds?

Calm down, I told myself. Women have been having babies since time began. They instinctively know how to do it. It's just like passing a bowling ball, nothing to it. She just needs to be reassured. So I took her hand. 'Everything's going to be all right,' I said, calmly. But my mind was racing. Could a PhD be struck off for malpractice?

Ten minutes later the attendant returned with a nurse.

'Thank you, nurse,' I said, as if talking to one of my staff. 'I'll just leave it to you then.'

'I think you'd better,' she said icily.

'Thank you so much, doctor,' said the mother-to-be. 'You were so calm. I can't tell you what a difference it made.'

That night I slept uneasily as the train raced northwards into the night and my mind raced into dark tunnels where in the echoing distance I could hear a baby crying.

Returns

Jack either liked you or he didn't. Even though I had almost sunk his favourite boat in the Rapids and encouraged John to pass water in public, he now proffered an invitation: 'Trevor, I would very much like you to collaborate on all our future projects as a senior member of the party.'

I was, of course, delighted. It was a wonderful opportunity. I had already decided that part of my future research would be on the interactions between grazers and their food plants, so the team projects at Lough Ine would complement it perfectly. Where better to do ecological field research, and who better to do it with than Jack and John?

So in 1967 I took what would become an annual journey to Dublin and then south. As usual, the tour of the Guinness brewery took so long I missed out on the free tasting. It was the guide's fault, he lingered on the trivia of malt and yeasts: 'And,

believe it or not, the yeast used for fermenting lager is called
Saccharomyces carlsbergensis . . .'

I know, I know, I longed to shout. Get on with it, I've a train
to catch!

The previous year the letters section of the newspaper had
been full of demands to remove Lord Nelson from his column in
Dublin. He was suspected of being an English spy, keeping his eye
on the Irish. In 1966 it was obligingly blown to smithereens by
nationalists.

'What happened to Nelson?' I asked.

'He left suddenly,' I was told. 'By air.'

So as not to outstay my welcome, I too left by train for
Cork.

Beside the bus station in Cork the temperance movement was
gaining followers – now three inebriates slumped against the
statue of its founder. Cork has always had trouble with statues.
Long ago, Lord Chatham was given the freedom of the city and a
fine monument was erected, but later relations soured. 'Cork,'
bellowed his Lordship, 'that refuge for pirates, that spawning
ground for smugglers . . .' So the city fathers removed his statue.
A wooden effigy of the Catholic James II was put up, but after his
defeat at the Battle of the Boyne it was decapitated and restored
with a new head in the likeness of the Protestant William of
Orange. George II warranted a figure on horseback, but it lurched
forward at such a dramatic angle that crutches had to be placed
beneath the rider's arm and the horse's belly. 'As might be
imagined,' said a contemporary guidebook, 'the effect is most
ludicrous.'

That year, Jack's party had sailed overnight from Fishguard
to Cork. Their travel arrangements had, as always, been made by

John who ensured that everything went smoothly. The only hitch came when one student retired to his cabin and found it occupied by a stranger. The steward challenged her through the locked door but she claimed to have booked it weeks ago. 'What's your name?' he asked, but she merely shouted, 'Weeks! I booked it weeks ago!' It took half an hour to establish that her name was Miss Weeks.

'Had she been a wee bit younger,' mused the steward, 'you could have come to an amicable agreement.'

When they arrived in Skibbereen, Jack got waylaid while the rest of the party waited and waited and waited at the Eldon Hotel. Eventually they began lunch without him. Jack arrived to a bowl of cool soup generously laced with cayenne pepper. To their disappointment he slurped the lot without flinching. Jack was truly impressive in adversity.

I came independently and, thanks to Jack's recommendations to the Royal Society and Royal Irish Academy, was flush enough to fly to Dublin. The pilot announced the estimated time of arrival as 19.27, adding, 'that's the hour, not the year'.

I had entrusted a backpack to the airline with my usual pessimism, 'Can I have a ticket to wherever you're sending my luggage?' Served me right for being so smart; it didn't arrive. My sleeping bag, toothbrush and jeans vanished without trace on a direct, forty-minute flight across the Belfast Triangle.

Aer Lingus was obviously used to this sort of thing: 'Just buy what you need and send us the bill.'

Arriving late, I had to spend the night in Skibbereen.

'I've arranged just the place for you to stay,' said Mrs Donelan. 'Only one pound seventy pence a night, with breakfast thrown in. Go up the street, round the corner and it's three doors this way. Ask for Mrs B*******.' It sounded like Mrs Brisket or Broccoli, something edible.

'How do you spell it?' I asked.

'Don't you be worrying how to spell it. When you find the house, she's the only landlady there.'

'What number did you say?'

'Oh, forty-something I'm thinking. But you won't be needing the number, it's the one with the white door.'

So I trudged along a street of white doors and sure enough at number fifteen there was Mrs B. She was a jolly, plump woman with something wondrously mobile beneath her pinny.

'And have you eaten?' were her first words.

'I snatched a sandwich in Cork.'

'A sandwich,' she cried in horror as if I'd said a frog and a couple of slugs. 'That's no food for a growing man. Sit yourself down and I'll see what I have.'

In the parlour I joined the only other guest, a cadaverous man who merely nodded and returned to expending a entire box of matches trying to light his pipe.

Mrs B brought in a hillock of cakes and a pot of tea on a silver-plated tray with a lace doily. 'The boy's starvin',' she informed the smoking corpse. 'Mr Hurley travels in ladies' undergarments,' she told me confidentially. Somehow, I merely smiled politely.

The parlour was heavy with dusty drapes and a smoke-yellowed glass chandelier was dragging down a ceiling held together with sticky tape. 'The house was never much,' said Mrs B sadly, 'but now it's not what it was.'

The walls bore a flight of jugs arranged in decreasing size, and heavily embossed plaster plaques, one with a lurching galleon and the other a water wheel as seen by the eye of a submerged trout. There was also a framed roll of honour of the IRA volunteers killed in 1920, and a copy of the declaration of the Provisional Government of 1916:

> We declare the right of the people of Ireland to the ownership of Ireland . . . The long usurpation of that right by a foreign people and government has not extinguished that right, nor can it ever be extinguished except by the destruction of the Irish people.

I felt uncomfortably British, and retreated into the black velveteen cushions painted with scenes of old Ireland in iridescent green and gold.

I paid Mrs B with wilted bank notes that were almost transparent. All the banks in Ireland had been on strike for nearly three months and the notes were worn out, yet somehow the economy clanked on.

Deathly Mr Hurley, surely an uncommercial traveller, left without a word and went to haunt another region of the house. He was replaced by Breeda, Mrs B's buxom daughter who sat staring at me with great cow's eyes set in a field of freckles. In those days everywhere I went there was a landlady's daughter. She smiled a lot and leaned towards me, revealing a deep divide and a tangle of straps and scaffolding. I wondered what it was in a seventeen-year-old that needed so much restraining.

I decided that desertion was the better part of valour: 'I'd better go up and have a bath.'

'I'm feelin' a little dirty myself,' she said with a sweet smile.

Fortunately the bathroom door had a huge mortice lock. Thus secured, I turned to find I was locked in a cell from the Château d'If. The window pane was obscured with fuzzy plastic sheeting so that even a twelve-foot tall peeping Tom couldn't look in, and the lavatory cistern flushed continuously, imparting an eerie mistiness to the room. The toilet paper was soggy and the wallpaper had peeled from the walls to reveal it was only masquerading as ceramic tiles. It had been applied not in rolls, but in a hundred irregular pieces, a masterpiece of the paster's art. The shower curtain displayed a mother's touch – the angel fish were upside down. I drew it aside to reveal a vibraphone of pipes, but no shower. Instead, there was an ancient bath that could be aged like a tree by a series of ginger rings. The 'hot' tap dripped frigid water into a jam jar slung beneath it on a string. I decided to undress as little as possible and just wash the most visible bits of my anatomy. Dirt, I reasoned, was less dangerous than typhoid.

My bedroom was the smallest I had ever occupied with just enough space to edge between the bed and a wardrobe with knobs as big as tennis balls. I tried to close the ancient Venetian blinds by twisting a ball on a rod until it fell off.

On the wall was a small framed cutting, a 'Thanksgiving' notice once placed by Mrs B in the *Cork Examiner.*

> Thanks to the dear Sacred Heart of Jesus through the merits of His Five Precious Wounds, and to the blessed Virgin Mary, blessed Oliver Plunket* and the other Saints for great favours received – asking more.

Beside it was a conventional prayer to Mary that ended unexpectedly with a plea for us to be preserved from 'things that go bump in the night'.

Mrs B had assured me that the room had been blessed before I arrived, and would, no doubt, be exorcised after my departure. I undressed beneath a looming wooden crucifix and fell into the candlewick caress of the counterpane. The marshmallow mattress enfolded me. On concentric lines of damp on the ceiling someone had written... 1959, 1960, 1961 . . . I sank into sleep and dreamed of landladies' daughters who went bump in the night.

The next morning I tried to replenish my missing clothes in Skibbereen. It was a mistake. The draper's shop was a Victorian emporium with shelves laden with rolls of dusty lace and oilcloth, pinafores, caps and cardigans. Nothing had changed for a hundred years except for the stock . . . maybe.

A little old man protruded above the mahogany counter, with eyes a-twinkle and a loveable leer.

'A pair of tennis shoes, three T-shirts – you only have brown? – make that two. And a pair of jeans please.'

The wee man piled the order on to the counter. This was

---

* The last Catholic martyr – executed at Tyburn in 1681.

long before designer jeans and clearly no one had designed this pair. I explained my predicament.

'Begob, that's different entirely. What else will you be needing? Let me see now . . .'

How could I tell him there was nothing else in his shop I could possibly want?

He licked an indelible pencil with his purple tongue and began to list things: 'Two pairs of trousers, an Aran sweater, one pair of Sunday boots, a dozen linen handkerchiefs, a gentleman's umbrella . . .'

'No, really, I don't need all those things.'

'Ach there's no need to buy them, just send the bill to Aer Lingus. They've pots o' money.'

The previous draper had also been a JP and, so the story goes, it was not uncommon for defendants to flash the lining of their coat to reveal his label. With his livelihood staring him in the face, he often erred on the side of leniency. Is it any wonder that County Cork is considered God's own country, housing the devil's own people?

I was fed up with walking the five miles from Skibbereen to the lough so Ernie Donelan took me in his antique taxi, now part of the fleet owned by:

Ernest Donelan & Son

Complete Funeral Furnishers

and

Car Hire Service

Personal supervision, Caskets, Coffins, Wreaths, Habits

The journey was as uneventful as any trip could be with Ernie at the wheel. He overtook a tractor by forcing it into a ditch. At one point we got stuck behind a cart and Ernie pummelled the horn. The farmer turned slowly, touched his cap and said, 'Is it a pain in your belly you're havin'?'

Ernie loved to tell stories while he drove, and politeness dictated that he always looked at you when he spoke. This once caused him to swerve into the Donovans' farmyard. 'Oh God,' cried Mrs Donovan. 'Is that the priest come to call again?'

When we reached the lough-side road, steam began to seep from beneath the dashboard and a warm gravy dripped on to the carpet. Ernie must have thought a mist had swept in, for he turned on the windscreen wipers. Fortunately we had reached the postbox, where I got out.

'You seem to have a sprung a leak,' I ventured politely, as the cloud thickened and the car floor submerged.

'Not at all,' said Ernie. "Tis only a bit of steam.' So, as the brim of his grey trilby went limp, he shot off down the road with plumes of vapour billowing from either side of the car as if it were rocket-assisted.

My backpack did not turn up until after I had returned home again. A limousine arrived at my front door. 'This yours?' said the chauffeur proffering a musty rucksack.

'Where on earth has it been all this time?'

'Dunno. It was found in the hold of this morning's plane from Dublin.'

It had been shuttling back and forth to Dublin for six weeks.

Cave

The first joint project on which I worked with Jack and John was a study of a cave outside the lough.

I navigated the rowing boat down the Rapids at slack water and out along Barloge to Bullock Island. The island is joined to Dromadoon by a bank of shingle called the Coosh. The name comes from the Gaelic *cuavas* meaning stepping stones, but *cooosh* is also the sound the waves make as they rush through the pebbles.

The island crouches at the side of Barloge Creek like a great animal resting after a meal, its dark mouth swallowing the waves. For 350 million years the sea had insinuated itself into every crevice, forcing the slates apart and plucking them away. Now the cave's entrance is five metres wide and nine high, although at least three metres are always submerged; from Barloge the cave resembles a stage set with stiff curtains hanging from the ceiling.

Even though it is not facing the open ocean, on rough days waves surge into the cavern to shout at the darkness. On the highest spring tides the swell can carry an incautious boat inward and upward and the huge blades of slate in the roof could guillotine the passengers. But on calm days you can ease a boat into the gloom and follow the blind waves as they feel their way along the walls.

The cave bores its way into the rock for over ninety-six metres, straight as an engineer's tunnel. All the zones of animals and plants rise diagonally up the walls following the swell of the waves as they funnel in. At the cave mouth, sun-shafts spotlight the swaying kelp, but as you venture further in, warding off the narrowing walls with the butt of an oar, the shadows close in and the kelp vanishes. Here the walls are hung with the lank hair of maroon seaweed, and a species called *lucifuga* hides from the light. These become progressively stunted inside the cave and abandon their regimented pattern of growth as if they have lost their way in the dark. Further still, and even these give way to the shade-tolerant crusts of pink coralline algae and the great grey mounds of the elephant hide sponge. The cave mouth recedes to a tiny bright arch in the distance and in the darkness there are only blind invertebrates and me.

It was this zonation that drew us to the cave, for it is almost identical to that found by a diver as he descends. However, the conditions are different. As you sink deeper into the sea, the light gets dimmer, but it also changes colour because water absorbs light selectively; six metres down all the red light has gone and a cut appears to bleed black smoke. In the cave there was the same dimming of the light but without these changes in its quality. It was one of Jack's 'natural experiments'. Just as he had used the Rapids as a system that demonstrated the effects of water flow, the cave tested which properties of light most influenced the plants. We already knew that different seaweeds have different light-sensitive pigments which collaborate in photosynthesis, and there were several theories to explain how these might influence the distribution of the plants. Here was a chance to put them to the test.

In 1967 we plotted the distribution of all the organisms in the cave, then set up photocells to measure light of different wavelengths at the inner limits of various species. Readings were made every fifteen minutes from dawn to dusk over several days, to cover different states of tide. This would have been a dismal job in the dank and dripping cave, but the long leads allowed the control panel to be sited on top of the island. This year Win had joined me at Lough Ine, and we spent many wonderful hours alone in the glorious sunshine, with the deserted Tranabo Cove to the east, empty Barloge to the west, and the Atlantic Ocean before us. It was our island at the end of the world. The view was breathtaking and, had anyone viewed us, it might have taken their breath away too. But we never missed a reading. The next summer we would be married.

When the work was finished, the photocells had to be recovered. Jack had planned to do it the next day, but the weather began to deteriorate. 'The wind is getting up in the west,' he said.

'So it is,' I replied.

'It looks as if it will get much worse before morning. I'm concerned for the photocells in the cave.'

'Do you want to go get them?'

'I think we should.'

So we slipped down the Rapids in the watery moonlight. It was almost midnight when we entered the cave. We eased into the flawless dark, but behind us our wake stimulated microscopic plankton to luminesce. As we pulled up the light cells, their tethering ropes lit up like iridescent green worms. A great gob of phosphorescence glowed on the blade of the oar, but a torch revealed nothing.

One of the lines was adrift.

'It has come loose,' said Jack dejectedly.

'So it has,' I agreed.

'It's way out of reach. We may lose it in the storm.'

'Would you like me to dive down and get it?'

'Without your gear? Would you?'

I stripped off, slipped over the side into the chill water and retrieved the apparatus. Underwater the outline of my body was etched in green sparks.

'Trevor recovered the missing photocell,' Jack wrote years later, 'and came up glowing from head to foot in brilliant phosphorescence – very appropriate.'

What he said at the time was: 'If only I had a scalpel, I could scrape off a sample.'

In spite of my green glow I shivered all the way home.

The results were later published, and set the pattern for our future joint efforts. I wrote the first draft of the paper, Jack calculated all the data and John did the illustrations.

Then Jack surprised me. 'How would you like to present the results for us at an international symposium?' he asked.

I had intellectual ignition.

Someone had planted sapling trees on the leeward side of Bullock Island. They were having a hard time getting established, but we never saw any sign of the owner – which was probably just as well as we were working there without his permission.

Sixteen years later Win and I moved to the Isle of Man

where we met a charming Irish woman called Penny. She and her husband Brian came to dinner and it transpired they had lived in West Cork, not far from Baltimore.

'In fact we still own a little patch of land on the coast there,' said Brian.

'Whereabouts? We know the area well.'

'It's only tiny,' he said modestly. 'You won't have heard of it. It's called Bullock Island.'

Awakenings

As a senior member of the party I could no longer just do my own thing at Lough Ine, I had to collaborate in the joint research.

At the lough, 'Lieutenant' Norton became responsible for numerous individual research tasks and ensuring that platoons of students knew what had to be done and why, as well as impressing upon them the rigour with which tasks had to be completed. In science, careless is useless. But perhaps my major role was to bring enjoyment to even the most mundane or laborious tasks. The official expedition diary written by the students reported that when processing collections of sea-floor mud, 'Trevor was a great help and actually made sieving those samples fun! We didn't know it could be done.'

Over the winter, back in Glasgow, I would spend long hours identifying any samples of algae that had been collected on the previous expedition, and helping to plan the research programme for the coming summer. This involved lengthy correspondence with Jack. Unfortunately, he often wrote by hand and his writing was small, neat and totally illegible. I think he used a different alphabet from the rest of us – Cyrillic, Erse, Ogham? To make matters worse, he often failed to separate the words, so that each line read like a Welsh signpost. By comparison, the decipherment of Linear B was a doddle. The West Cork writers, Somerville and Ross, inadvertently captured it exactly in one of their *Reminiscences of an Irish RM*: 'No individual word was decipherable, but, with a bold reader, groups could be made to conform to a scheme based on probabilities.'

The first task was to determine where the breaks between the words might lie. It was like searching for khaki golf balls in deep rough. I resorted to writing a tentative translation beneath each line, like a schoolboy with his first French primer, and then replied based on my best-guess version of his letter. If, as he must surely have done, Jack sometimes received a response that bore no relation to his original letter, he never enquired whether I was slipping into dementia.

It wasn't just me, no one could decipher Jack's hand. One of his research students submitted a draft chapter of his thesis and Jack returned it with a single scribbled comment. It was illegible, so he asked Jack to translate. It said, 'I cannot read your writing.'

Curiously, Jack's numbers were perfect. For every scientific job the raw data were written up on large record cards. Task 1000, *Dog whelk collections from Landing Gully*, had been completed in 1965. Jack's tables of numbers were immaculate, he could have been an emanuensis for a mathematician. They would have been examples of best practice – if only we could have deciphered the headings.

The other senior member of the group who had to wrestle

with Jack's writing was Louise Muntz, who had first come to the lough in 1959 when she was his student at Bristol. She was now a lecturer at Reading University, and was a practical person who could be relied upon to get the job done and do it well. One night the macho male students exhibited their strength by pleating metal bottle caps between their fingers. Louise viewed their efforts, then revealed a cap that she had folded completely in two. She tossed it on to the table with a single word – 'Better!'

There was also Viv Pratt, a student from Bristol who went with Jack to the University of East Anglia to study the behaviour of striped snails known as top-shells. Legend has it that when Saint Patrick was pursued he hid in a top-shell using his cap to close the 'mouth', but his enemies found out and he had to cap every shell to fool them – so now every top-shell has a lid. As the opening of the shell is only a centimetre or so across, I assume that Saint Patrick was not a very big fellow.

Top-shells are the best barbers in the sea. They keep the bigger seaweeds clean by shaving off the stubble of small algae that foul their surface and would impair their photosynthesis. The snails wander on to the tops of rocks at dawn and retreat down below at dusk.

Viv set up a large glass-bottomed aquarium tank full of stones in the Dromadoon lab so that she could see what the snails got up to beneath the rocks. She monitored them throughout the night and I kept her company and worked through some samples of algae while she slept. Every ten minutes the alarm went off and Viv got up, viewed the snails through infrared light (so as not to interrupt the dark), noted how many were below the rocks and active, and climbed back on to the camp bed.

'Want a cup of coffee?' I asked. But she was already asleep.

I wondered whether she was waking up at all or just rising on auto-pilot, scribbling down some arbitrary numbers and crashing back to bed, so I checked her readings. They were spot-on.

Viv also devised means of studying snails in the lough. When they were moved to below their normal zone on the shore, they resolutely trudged back. The snail's brain, such as it is, ignored all the obvious cues such as slope and gravity, or the position of the sun, and responded to the relative darkness of the upper shore which loomed above it on one side. When Viv erected a dark plastic wall on the downhill side, the snails set off in the wrong direction.

She also suspended floodlights on the quay, fed by a generator. Their purpose was to fool the snails into thinking it was daytime during the night – and perhaps to electrocute the experimenters. During the day she enclosed a bit of the shore in a black plastic 'coffin' to keep out the sun. We ensured it was light-proof by putting Viv inside to look for leaks.

The behaviour and reproduction of many marine plants and animals is governed by internal rhythms, which synchronise with the phases of the moon (and thus the tides), or with the duration of the night. Viv showed that the snails' rhythm was entrained to coincide with daybreak and dusk, just as our own internal rhythm naturally runs on a shorter day than the twenty-four-hour clock, but close enough for our patterns of activity and sleep to be fitted into the fluctuations of night and day. Jack, however, was the single exception. He was constrained by nothing but work and the tides. While Viv was disrupting the rhythms of activity and rest in snails, Jack, on principle, was disrupting ours. His enthusiasm for all-night projects wore us out and, being only half awake, we were not the most acute observers. For example, we knew that the purple sea urchin was abundant and obvious in the shallows during the day, but absent at night having crawled down between the rocks. Observations over twenty-four hours would, Jack reckoned, establish exactly when the animals moved. Every half-hour three students rowed out to the observation site armed with view boxes and torches. Jack had even invented special little waterproof torches on long rods that could be manoeuvred into crevices underwater to illuminate the cowering urchins. But they weren't lights, they were darks; if switched on in daylight they could flood a room with darkness. They were given the 'Order of Renouf', an occasional award for the most useless piece of equipment. Other winners had been the unsound echo-sounder and the intemperate temperature recorders.

Next morning we were surprised to find that the urchins had moved intermittently in sudden leaps during the night – whenever the students changed shifts.

By the end of a long hard day at sea in a small boat I was ready for bed, but when I shut my eyes the safari bed adopted the queasy sway of the ocean. Even so, I was soon unconcious.

Sleep came easily, waking up was the problem. Fortunately, we had two of the best alarm clocks in all Ireland – Jack and John. Jack, always first to bed and therefore first up, gallumphed around in wellington boots two sizes too big, in which his bare feet sounded like fat waves slapping into a cave.

Then there would be the dawn discussion with John. Last night's 'What about tomorrow's programme?' had become 'What shall we do today?'

'What's the tide doing?' Jack cried, striding to the tide tables pinned up on the door. 'High tide at Cobh is . . . add forty minutes for spring tide . . . an hour for Summer Time . . . three and a half for Low Slack in the lough . . .'

Then they discussed the rising wind and impending rain in such deafening whispers that soon we were all awake. Jack's voice was like no other. It had the strained boom of a lonely bittern.

Jack relished raising the dead. The only problem he had was when only one of the three occupants of a tent had an early call. Shrouded in sleeping bags like pupae cocooned against the cold, Jack was forced to admit, 'all women look alike in bed', a comment so out of character that Jack himself was taken by surprise, and smiled sheepishly.

He liked to startle people awake. If it was still dark, he would flash a torch in their face and shout, 'Fire!' Once he shook me and shouted, 'Quickly, Hold this!' I shot bolt upright in bed, then began to slump forward like a dog nodding before the fire. As I rallied, I realised I was white knuckled from gripping a rope. I forced myself awake and emerged from the tent to see the rope trailing away down the site. Bare-footed and in my pyjamas I staggered through the dew-drenched grass down to the quay and found that the rope was attached to a boat floating out in the lough. Another time Jack tried a similar trick, but this time gave the other end of the line to someone in a different tent, then shouted, 'Pull!'

Jack preferred the noisy methods. John kept coathangers

dangling in his tent to hang shirts on. When he wasn't looking, Jack attached a string to the wire hangers and threaded it out through the tent's vent so that a couple of tugs at dawn would set them jangling. He also inflated and tied a plastic bag to make a balloon and then jumped on it to make it bang, but in his enthusiasm he leapt too high and knocked himself out on a low beam.

Surprisingly, he never used the most effective alarm of all. One of the research projects required us to take simultaneous measurements of water temperature at several sites around the lough to see how it influenced the hydrography. To synchronise the readings, Jack bought an aerosol foghorn. Even at a distance the blast made your teeth tremble and depressed your eyeballs. It would have awoken the campers all right, and blown down the tents too.

The next best thing was the 'Irish warthog'. He would pour a little water into a plastic bottle which he then inserted beneath the tent flap. When squeezed rhythmically it belched loudly. If that didn't work, he flattened cans with a spade.

One student, Elsie, had a novel way of avoiding rude awakenings – she never slept. Every night, no matter how late we retired, there was Elsie sitting at the table reading or playing patience; next morning, no matter how early we rose, she was still there turning cards. To the best of my knowledge she slept only once in two weeks. But it was for twenty-four hours. I think this regime took its toll, for Elsie was peculiarly vulnerable to gravity. She could get one leg into a boat, but rarely two. Once, with one leg aboard she leaned back and the sea poured in. 'Trevor,' said Jack calmly, 'I'm all right, but I think your end of the boat is sinking.'

We were having trouble with rats. They had gnawed the

base of the mess hut door, and a couple of students had seen them scampering across the fields. In desperation Jack had written to the Infestation Control Division of the Ministry of Agriculture and received a *Handbook of Rat and Mouse Control for the Training of Rodent Control Operatives*. With immense ceremony and elaborate precautions, Jack laid bait: bread with poisoned peanut butter on it — even more poisonous than usual. But the next night I distinctly heard a scuttle of tiny claws outside the men's washroom. I wasn't afraid of rats (although I once knew a researcher of rodents who was terrified of them), but neither was I thrilled by the idea that they might nibble my nose in the night, and my low camp bed was well within rat-hopping range. I suppose they must have been on my mind when I went to sleep.

Next morning there was an early start, and Jack devised a novel alarm call. He fixed a large tin funnel to a length of hosepipe. The plan was to slip the improvised trumpet under the skirt of my tent, push it towards my ear and blow a loud reveille.

It was six o'clock when I began to surface, and something was scratching at the wall of the tent, trying to get in. Rats! Only half awake, I grabbed the nearest weapon — a large wooden mallet for whacking in tent pegs — and launched a frenzied attack on the unseen enemy, blow after blow, until I heard a groan. Rats squeak, I reasoned. This was a very deep squeak.

Jack lay outside on the ground, stunned, clutching his head and a sad, flattened funnel. He never again used that particular method for rousing me, for I was terrible when roused.

News of my stout repulsion of intruders spread, perhaps even to the rats, for they never bothered us again that summer. And nor did Jack.

Potheen

In 1968 John damaged a tendon in his finger. The local doctor encased it in a splint and bandaged it to resemble a mummified sausage. It stuck out – well, like a sore thumb.

He felt sorry for himself and slumped in a chair impersonating a stranded jellyfish. Then a letter arrived that cheered him up – he had been appointed to a professorship at Sheffield University. I heard the news from Jack, and we decided to hold a surprise party on Sunday.

When I met John, he was beaming. 'A personal chair,' he said, adopting an expression he assumed might convey modesty.

'That's marvellous, John. Now about these worms . . .'

His face crumpled with disappointment.

On Sunday evening the Bohanes came over for our usual soirée. John had recovered sufficiently to assail them with tales of

Sheffield city councillors. 'At one council meeting it was suggested that a gondola might look nice on the lake in Crookes Valley Park. They all agreed, then Councillor Oglethorpe got carried away and said, "Let's get two and breed t'buggers."'

Then he reminisced about his early days in the zoology department at Hull University. 'They were all batty, quite batty. The Prof cycled to work each day in full academic robes with a parrot on his shoulder and kept the results of his wife's miscarriages pickled in bottles on the laboratory shelf. After two years I moved to the comparative sanity of Zoology at Sheffield, but even there I had to share an office with a colleague and her pet python which slithered loose around the room.'

The Bohanes had brought a bottle of potheen – the spirit made from fermented potatoes. After all, why eat a potato if you can drink it? Every shed and barn was abubble with chemistry in those days. That very week I had seen the *Garda* sneaking round the hedges on all fours hoping to surprise a local in mid-experiment. The turf was stiff with hidden bottles. If you fell in a bog you'd sooner poke your eye out than drown.

I had never had anything like it before. It smelled and tasted like lighter fuel. My gums shrivelled away and I was sure my teeth had begun to dissolve.

'Take care,' warned John Bohane. 'It'll slip down like a torchlight procession and melt the nails in your shoes. Take a nip too many and *if* you wake up in the morning and it's sober you're feeling, take a swig o' water and you'll be runnin' across the fields in your pyjamas.'

'It's true,' said Neilly. 'Sure the stuff would make a horse out of a cow.' He told us of a local who drank it by the pint, then was laid up for a fortnight close to death. 'It was his regular routine: two weeks on the potheen and two weeks off ill. So I asks him,

"If it's sick it's makin' you, why would you be doing it?" "Ach," says he, "when I stop the drinkin' I feel so grand I just have to go out and celebrate".'

The Irish have always had an ambiguous attitude towards temperance, but they don't deny that liquor causes anger and irrationality – especially in abstainers. When a well-known Gaelic scholar was invited up from the country to address a conference in Cork on the influence of temperance on Irish culture, he was nervous about facing a large audience, but after a large slug of potheen his lecture went just fine.

'My first taste of potheen was in 1938,' John Ebling confessed. 'We had decided to make a big batch of toffee. So I asked the shopkeeper for ten one-pound tins of treacle. He couldn't have been more shocked if I had asked for ten Lee Enfields.'

'"Oh I can't be selling you that. 'Tis against the law." "Treacle is illegal?" I replied. "I believe it can be used for the manufacture of potheen," he said, as if he had heard a vague rumour to that effect. "But I tell you what I can do for a gentleman like yourself. I'll wrap them tins in brown paper and if anyone asks, say it's paint you have." Then he hugged himself and tittered as if it were a great joke.

'As we were leaving the shop he beckoned me to one side and invited us back that evening for a little get-together. We had to come by the back alley and use a secret knock. It wasn't much of a secret, for when we were let in, we found half the town was there swigging potheen. Fortunately, there were only two *gardaí* in Skibb in those days and one was off duty, drinking in the corner. There was a rare sing-song and I think I can remember one of the tunes . . .'

Before the groans had subsided John began to croak a jig:

On yonder little hill,
There's a darlin' little still,
Its smoke curling up to the sky.
And it's easy to tell
By the whiff and the smell,
That there's potheen, me boys, close by.
As home we roll.
We'll drink a bowl,
or a bucket of the Mountain Dew.

To forestall the appearance of the song books, the official celebration of John's promotion began. I entered in an improvised academic gown with a sou'wester moulded into a cap, preceded by a mace bearer and followed by a student carrying a cushion that held all the ceremonial objects John would need in his new post. He was presented with:

A razor for splitting hairs.
A toy trumpet of his own – to blow.
A spanner for the works.
A roll of red tape.
A coil of rope – one end for swinging the lead, the other for spinning a line.
A ceremonial finger-stall, sequinned and be-ribboned.

As John donned the finger-stall, everyone else placed a hand on the table to reveal that they each had an immense sausage finger swathed in bandage.

Finally, John was dressed in a mortar board and given a privy seal and chain of office. Jack made a speech and presented him with a congratulatory scroll – beautifully calligraphed by Liz, one of the students. I then sang my latest composition:

> At Sheffield a chair was vacant,
> But supplies of talent they were low.
> If I say his jokes were blatant,
> The man they've chosen you may know.
> Oh, not John, not John, not John.
> Afraid so . . .

> A DSc from Bristol City,
> For a thesis on the pituitary gland.
> A flimsy thing, abstruse and witty,
> That no one else could understand.
> Oh, what rot, John, rot John, rot John.
> We know . . .

> The senate hailed his dissertation,
> Which for a thousand pages ran.
> But he gives me constipation,
> when he sings 'The Music Man'.
> Oh no John, no John, no John . . .
> No!

Two students, Liz and Cathy, brought in a cake they had baked and decorated with a Baltimore beacon fashioned from a candle and a glacé cherry, and we all had Irish coffee and a marvellous time.

Professor Ebling would attend many ceremonies in the years ahead, but few that he would enjoy more than the evening he wore a ceremonial lavatory chain and we sang a slanderous song about him.

That evening John was on transmit rather than receive and held centre stage, boisterously telling tale after tale. Jack was quiet and thoughtful. His former pupil was now officially his peer. For all his investment at Lough Ine, Jack had earned the respect of the locals, but John was the one they loved. I wondered how he felt about that.

Jack was always generous with young scientists like me, and immediately forgave me for attempting to batter him to death. But he mentally copyrighted his colleagues, and was prickly when anything appeared to undermine his notion of the team project. When anyone who had worked at Lough Ine considered publishing independently, they were quickly disabused of the notion.

John Harvey, a colleague of Jack's from the University of East Anglia, studied the hydrography of the lough in 1969. He sent a copy of his findings to Jack and, when invited to present a paper at a scientific meeting, decided to talk about his Lough Ine data. He was summoned to Jack's room.

'You were invited to Lough Ine and given facilities on the understanding that this was part of a collaborative venture, and could not be published without the expressed permission of the others. It is most improper that you should have chosen to flout these strictures.'

'I'm very sorry, Jack, but I wasn't aware any conditions were placed on my work. I very much regret the misunderstanding.'

'Quite so. It would be most un-colleaguelike to present your results independently at the meeting.'

'All right, if that's how you feel. I'll go away and think it over.'

At this, Jack exploded. 'You cannot possibly continue to be a colleague if you even consider giving the paper.'

'Are you telling me that you won't permit me to talk in public about my own results?'

'Certainly. You accepted my facilities on those terms, and you must abide by them.'

Harvey withdrew his paper, but never returned to Lough Ine, so we lost the invaluable services of an oceanographer. And it was all for nothing, for only the briefest abstract would have been published from the meeting, and this would have in no way compromised the publication of the work in full.

I knew nothing of this at the time, and later fell into the same mire. We had been studying the underwater algal vegetation, with the bulk of the work being done by me and a diving team, led by Tony Larkum. The initial survey was completed and Tony was anxious to publish, but, getting no response from Jack, he wrote to me: 'I am mystified as to what is going on. As you know, I worked up all the results a year ago and wrote down a lot of my ideas, which I sent Jack for comments. Have heard nothing for six months . .'

Tony delivered an account of the work at a meeting of a learned society, and kindly included my name as joint author. Again, only a tiny abstract would be published, but Jack was seriously displeased – although it didn't goad him into publishing the full results for another twenty years – an eternity in the career of a young scientist.

Stores

At Lough Ine, Jack chafed if he wasn't busy doing science, while John easily adjusted to the rhythm of the place and had an instant rapport with the local people. Lacking John's conversational Vaseline, Jack perhaps resented his tendency to talk in exclamation marks and sometimes, instead of conversing *with* everybody, conversing *for* everybody. But Jack failed to appreciate the extent to which John smoothed things over – as Michael Sleigh once said, Jack made it happen, but John made it work.

John liked things to be orderly, whereas Jack tolerated a fair degree of chaos under the misapprehension that it was order, it having been ordered by him. One of the many things on which they disagreed was what equipment should be brought afresh each year and what should be left behind. There was much to be said for bringing almost everything that was needed for the job to ensure that it worked. But Jack had faith in storage, so we stored.

Some items were stowed so securely that they were never seen again. But the biggest problem was the damp. Salt-spattered equipment never dried and corrosion proceeded apace. Drying, cleaning and oiling before putting it away only slowed the process. After a winter's absence, we returned to find drawers full of instruments set in crusts of corrosion: callipers fixed agape, scissors solidly shut, scalpel blades fused together into a rust sandwich. One of our balances went arthritic and, from then on, everything large or small would weigh 13.52 grams. There were boxes of mouldy string, moth-eaten dusters and dried-up ballpoints. Wood suffered invasions of wet rot and woodworm, and although these plagues spared the wooden rulers, inexplicably someone had stolen all their markings and rendered them useless. Pencils fell into splinters the first time they were used – which was perhaps fortunate, for the green erasers had turned to bricks. Impatient rodents hastened the decay by chewing the pencils and making papier mâché of our record cards.

We had dozens of plastic bowls and baby baths for sorting samples, but over the winter they developed hair-line cracks, so the first time they were filled and lifted the handles fell off or the bottoms dropped out, drenching your jeans.

It was also a graveyard for garments, though, to be fair, they didn't always start off in prime condition. I listed the gear I had left behind in the mess hut attic the previous year:

> Diving suit – bootees ripped
> Wellies – left one leaks
> Two pairs of gym shoes – one with only half a sole
> One old towel – threadbare in parts
> One pair of jeans – not much good.

Unfortunately, the moths weren't snobs, they would eat anything.

Another problem was shrinkage – not of clothes, but equipment. The glass-bottomed view boxes became leaky and the boats let in light between the planks. They had to be soaked for days before they could be used safely.

There was also the 'selective shrinkage' of bits going missing. We were left with cages lacking lids, bowls without bottoms and string without strength. I never cease to wonder how all the gear stuffed into Tutankhamen's tomb was as good as new after three thousand years, whereas we couldn't persuade an eraser to overwinter.

A wooden stool furnished a winter picnic for woodworm. Jack soaked it in creosote to prolong its life, but it never dried, and every occupant developed an embarrassing brown bloom on the seat of their pants. Creosote kills woodworm, but doesn't fill in the holes. The stool's legs were hollow with burrows, and once, while I was bent over a microscope, they vanished in a puff of creosotic sawdust and I fell to the floor.

The attic store was where food went to die, so Jack decided to burn some old fish paste, some very old fish paste indeed. Prehistoric paste is not renowned for its inflammability, so he started the fire with the creosoted stool and an old rubber raincoat. I kept the students upwind so that nobody suffered a fatal inhalation of smoke.

Observers miles away assumed it was Ireland's first volcano, situated somewhere offshore, for it bore a distinct aroma of fetid fish. It was the blackest smoke I had ever seen . . . until five days later when Jack, still giddy with fumes, burned two old tents, four ground sheets, three welly boots and a pair of plimsolls.

The smoke pall would have done justice to a blaze in an oil refinery, but we didn't expect firemen to rush over — at the last fire in Skibbereen the crew from a town miles away had arrived *before* the local brigade, and crowds had lined the street to jeer them when they eventually turned up. Perhaps they had lost heart, for no fire in rural Cork broke out within a thousand hose-lengths of a hydrant.

A lot of stuff was stored in vast quantities. The previous year, Jack had ordered cement from Fullers hardware shop in Skibbereen, and two tons had arrived unannounced at John Bohane's house. The only place he had for forty large bags was the lean-to shed outside, and after a winter in the driving rain the cement turned to concrete.

Jack was furious; he had ordered two hundredweight, not two tons. Fullers would just have to take it back. And they did.

'But sure it'll be useless now entirely,' said John Bohane.

'Not at all,' the Fullers driver replied. 'We'll just hammer it into little lumps and put a bit in all the other bags.'

The only objects that Jack refused to store were Irish matches. He believed they were liable, indeed designed, to combust spontaneously and consume the entire building. I have no idea where he got this notion, as in my experience the wood invariably snapped long before you could generate enough friction to ignite the red tip.

'Make way! Make way!' Jack would shout, rushing out with a gross of boxes of Irish matches held out at arm's length. He condemned them to another brief but spectacular bonfire. Eccentrics do make for good stories, but can be hard to live with.

The loft was a treasure house of equipment and provisions. In winter the whole floor area was filled with folded tents and safari beds, the ground sheets slung over the cross beams. There was a crate of large, shining 'Easter eggs' – the class microscopes in their steel-domed boxes. These had been bought years before, because Jack thought that in their tin helmets they were 'ideal for dropping off Sonny Donovan's cart on the way down to the laboratory.'

Cupboards lined the walls of the loft, filled with boxes and the canned food left over from last year, plus almost anything else you can imagine. The boxes encapsulated the time; they burgeoned with Spam and Bisto gravy browning and Carnation evaporated milk. Every third box seemed to contain packs of candles, or balls of string, and there were sufficient tins of shoe polish to cover all the shoes that had ever existed in case they rose again and demanded to be cleaned. There was a packet of ballpoint pens bought in 1948 because they were advertised as being able to write when upside down or even underwater. They did neither.

Perhaps they are all still there awaiting rediscovery by an archaeologist. I can almost hear the explorer's hushed, reverential commentary: 'The heavy attic trapdoor creaked open, brushing aside curtains of cobwebs. I could taste the musty air, unbreathed since these treasures were laid to rest. As my eyes became accustomed to the ancient darkness, I began to discern weird dust-laden objects, wonderful things . . . and everywhere . . . the glint of string.'

Every box was numbered and its contents inventoried:

Box II: toothbrush holders
   fly spray
   birthday cake candle holders in [empty] custard tin
   wooden spoon [our highest award]
   song sheets [our lowest ebb]
   tin of Antiphlogistine [to reduce swelling]
   metal fly cover [last resort if Antiphlogistine didn't work?]

Anything corruptible was entombed in large metal canisters. To keep rust at bay, Jack insisted they were smeared with Vaseline. Unwittingly, he had invented the greatest rust accelerator known to man: the grease ensured that dust stuck to the canisters, the dust attracted moisture, and every speck acted as a focus for corrosion. Eventually, the canisters were breached and the contents spoiled. John discovered a suppurating mass of Jelly de Luxe, a whole tinfull of Boudoir biscuits turned to putty, and a hoard of

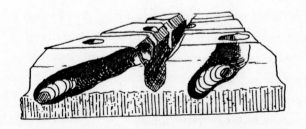

very old chocolate. Not only was its surface bloomed with a white powder, it was alive with squirming maggots.

'Spoilt,' said John, for the benefit of the short-sighted. 'The most unsavoury thing in the universe, except for the bluebottle.'

'Nonsense!' claimed Jack, the chocoholic. 'Just a few grubs. Perfectly palatable.'

Before we could argue he had bitten off a large chunk and was chewing it thoughtfully like a connoisseur at a cheese-tasting contest. After some time he swallowed and pronounced it fit for consumption.

'Not too bad,' he declared. 'Full of protein.'

But he didn't object when we threw the rest away.

On the first trip to Lough Ine in 1937, a great haunch of bacon had been delivered. One end was infested with maggots and had had to be discarded. When they'd complained to the shopkeeper she was appalled: ''Twas a wicked sin to throw away good bacon. Sure 'tis well known that the maggots only go for the sweetest part.'

The students returned home with a store of wonderful tales of their adventures beside the lough, so these safaris became famous back at the University of East Anglia.

Sadly, not all the undergraduates who applied could be accommodated, so Jack selected who could come and who could not. Jack's daughters had taken part several times in earlier years, and John asked whether his son might come too: 'Francis is free later in the summer and would love to join us at the lough.'

'I am sorry, it is quite out of the question,' Jack replied.

Perhaps if John had asked earlier, it might have been all right. But it was clear there would be no discussion and that Jack felt no compulsion to explain. John was hurt. I don't think it had occurred to him that Jack might refuse. Long ago, when Jack had bought his first patch of land at the lough, John had worried whether it would affect the nature of their collaboration. What had been implicit then was now clearly stated: Jack was not just the leader of the expeditions, he was also the proprietor of the land and the laboratories. He alone would decide who the other participants would be.

We were all guests at Jack's party, and would do well to remember it.

Evenings

In the long summer evenings, between washing up and final tick inspection, there was still light enough for a walk – and even if there wasn't, Jack, on principle, shunned a torch and set off anyway.

One night in the early years, malicious clouds covered the moon and the walkers lost their way and had to knock on the door of a cottage. After no response they knocked again, and from inside came the barking of dogs, the shriek of children and a man's voice imploring the Almighty to preserve them from evil. The door opened to reveal the entire family cowering behind father, who was brandishing a shillelagh. Goodness knows what they were expecting, but they were obviously relieved to see only bedraggled biologists, and escorted them down the track to the road. It was nearly midnight when they passed Mrs Donovan mending a fence in absolute darkness.

Getting directions was not always so easy. When asked where a road led, a local once replied, 'Sure it will take you anywhere in the world you'd wish to go.'

Jack preferred a brisk march to waft away the cobwebs. Trailing a short string of student volunteers, he would stride on scissor legs over to the Rapids through head-high bracken, bayoneted brambles and stinging nettles. Fifteen minutes and twenty ticks later he would reach the Rapids, sit for a while, then return, with a chorus of nightjars calling, 'Quick quick! Quick quick!'

Sometimes we tramped the evening lanes where walls of fuchsia bled petals on to the road. A succession of farmers went by on their carts. 'Grand day,' they all said, raising a finger in greeting. If it wasn't actually pouring ('flogging rain') or blowing a gale ('fierce terrible weather'), it had invariably been 'a grand day', never mind the gathering greyness in the west or the drizzle ambling towards us over the rim of the hill.

Another favourite summer walk was down to sleepy Baltimore. The village had been dozing just the same on another warm June evening in 1631 when two Algerian galleys slipped into the harbour and pirates sacked the town.

> The yell of 'Allah' breaks above the prayer, and
>   shriek, and roar.
> Oh, blessed God! The Algerine is lord of Baltimore.

One hundred and ten captives were taken into slavery and only two were ever seen again. For the next hundred years or so Moorish pirates were the scourge of the south coasts of both Ireland and England.

Later, Baltimore revived, then declined, as a succession of fisheries flourished and collapsed. The coast was once dotted with 'fish palaces' where pilchards were smoked, but the pilchard mysteriously vanished in the early 18th century. Then the mackerel and herring came under pressure – on a single day in 1786, three hundred French boats deployed their nets between Baltimore and Crookhaven. By the 1890s, women from all over Ireland gathered in Baltimore to gut and salt the herring. As late as 1925, 20,000 barrels of local mackerel were exported to the United States, but trade tariffs put an end to that.

Baltimore's only claim to fame is that it gave its name to Lord Baltimore who, in turn, presented it to Baltimore in Maryland. It is fatal for a town to give away its name, for often its ambitions are transferred too and the recipient does better than the donor. By the 1960s the population of Baltimore, Maryland, had topped 800,000 and was growing fast, whereas Baltimore, Cork, was still a village aspiring to be a yachty place. Already the draper's shop was masquerading as a bar and the post office was disguised as a boat yard. There was a rash of jolly visitors in Aran sweaters, little green wellies and sailors' caps that no sailor would be seen dead in.

If Jack didn't walk, he rowed, and often he would invite a couple of the students for an after-dinner trip around the lough; the brown boat was a natural circumnavigator. About an hour or so out Jack would reach into his back pocket and produce a couple of bent, warm bars of Cadbury's Bourneville chocolate for a picnic afloat.

As darkness fell and the moon rose, he might head for the mysterious foaming milkiness of the Rapids, but if it was overcast, he would take the boat beneath the overhanging trees on the western shore where in the darkness bleached sea lettuce hanging from the branches trailed across your face and the oars sparkled with soft emeralds of phosphorescence. In the flawless night he would glide the boat back to the invisible quay at Glannafeen without a bump.

One dark night I rowed over the deepest part of the western trough and let the boat drift. It was absolutely silent except for the dripping of water from the half-stowed oars. The beam of my torch scythed down through the water, and within seconds, dozens of pink worms wriggled into the spotlight. These were the emissaries of solitary, bottom-living animals that just for once swarmed together at the surface of the sea. The stimulus was, of course, sex. Males and females loaded sperm or eggs into their back ends then shed them like rocket modules on a mission. The moon orchestrated the proceedings so that all the worms spawned at the same time. The full moon fired the starter's gun, and it took the worms fourteen days to pack and ripen their rumps – so two weeks later, on the darkest night of the month, they swam upwards in search of a lost moon.

Unlike their sluggish parents, these ripe and ready young-sters were frenetic swimmers. The males circled the females in a nuptial tango. It was difficult to believe they were not entire worms. I put some in a bucket and they rushed around the edge like motorbikes on the wall of death. They were bursting to do something, and soon the water was clouded with sperm and eggs. After their brief dance in the dark, they died.

When I wanted to be alone, my favourite evening excursion was to Castle Island in the middle of the lough. It lay only 300 metres from the camp. Floodlit in the morning sun it looked close enough to touch, but in the evening it receded until it seemed too far to swim.

One evening I rowed out and beached the boat in Curlew Bay where the birds loitered on the beach, disconsolately probing the mud. Every afternoon, about half an hour before sunset, the curlews returned here for the night. In the early days the students would stand on a knoll beside the Rapids and call them home twenty minutes early. I sat on a rock in the calm and gazed at the tattered skeins of algae and litter of empty oyster shells in the shallows. On the point, the dark crucifix of a cormorant was drying its wings. I fell into a dream until there was a commotion in the bracken behind me — something approaching, louder and louder, closer and closer, a deer at least, a stampeding herd. Then suddenly an ungainly heron clattered out on its great heavy wings, just clearing my head.

I walked through the purple Michaelmas daisies to the ruins of Castle Cloghan (a stony place) on the eastern rise of the island. It was once the fortress of a great king called Labhraidh Loingseach. His original name was Maon, son of Ailill. When his father was slaughtered, he was forced to eat his heart and liver. Struck dumb by the experience, he escaped and became known as Loingseach — 'the exiled one'. He raised an army and returned to defeat his father's assassin and claim the throne of Ireland. Some say his speech was restored by the love of a beautiful princess, others by a blow from a hurley stick — the effect can be much the same. From then on he was called Labhraidh — 'he who speaks.' There is a legend, an old tale given an Irish accent, that states he was cursed with the ears of a horse, which understandably he kept under his hat. Only his barbers knew the secret, but not for long, for after each haircut he hanged the hairdresser. However, one barber was spared and whispered the secret to a plant. It may have been a reed that was later whittled into a penny-whistle by a passing minstrel, or perhaps a willow tree that was made into a harp. Either way the result was the same – it would play only one tune, and the melody had words: 'Two ears of a horse has Labhraidh Loingseach . . .'

Every rush and rowan tree and even the rocks around the lough took up the refrain. Louder and louder it became, until the king could stand it no longer and fled the castle, never to return, carelessly forgetting a crock of gold beneath the floor.

If there was gold below the flagstones, Fineen O'Driscoll never found it, for he died there destitute. Once, the O'Driscolls had ruled the entire coast from Kenmare to Kinsale and had become rich by collecting dues from fishing boats. But Fineen lost everything when *Dun na Seod* (Fort of Jewels) in Baltimore was razed in 1537 by Waterford men in reprisal for the seizure of one of their ships.

In the middle of the 19th century Bill Barrett, a local man, searched for the treasure one night. Before he had dug very far, he noticed a black cat outlined in the moonlight. It transformed into a huge dark dog with eyes aflame, as Irish phantom felines are wont to do. Bill stood terrified and the dog leapt into the lough, throwing up a wave almost big enough to drown him. He took over three months to recover his wits and never visited the island again. In 1872 the castle finally collapsed and entered its long retirement as a romantic ruin.

One evening a group of us trekked along the western shore road, through the tunnel of over-arching oaks, beeches and hollies, to the Western Quay. On the open road beyond we were bombarded by mussel shells dropped by birds high above us. How else could they crack them open? Each year hooded crows removed almost a quarter of all the mussels from the shores of the lough.

We decided to climb *Knockomagh* (the crooked hill), which broods over the north-west corner of the lough. We called it Lough Ine Mountain for, although only 195 metres high, it rose precipitously to make the most of its stature. Locals sometimes called it the Soldier's Hill, for long ago a young fusilier fell to his death when trying to steal a hawk's egg from its nest for his sweetheart, or perhaps it was because in the late 1850s large numbers of recruits to the Irish Revolutionary Brotherhood came up here to these lonely slopes to drill with their weapons.

A narrow trail zigzags up the steep, heavily wooded slope. It was hard going, but we were encouraged by intermittent views of the lough through breaks in the foliage. The summit was a grand rockery of outcrops and heather and, inexplicably, tiny spruce saplings that had been planted on the southern crest. Far below, the lough lay like a map, with the great blue Atlantic beyond. To the west, the River Ilen meandered away to the coast, and the hundred islands of Carbery were lost in the silver dazzle of the sea. We didn't wait for the sunset in case the path back was tricky in the dark.

We made it down more quickly than expected and decided to search for Saint Ina's well that nestles at the base of the hill. Just as the light faded, we found it hidden in a basket of overhanging hawthorns that had dozens of rosaries and pale, faded ribbons hanging eerily from almost every branch. The well is in a district called Pookeen (little Puck), for hereabouts pixies have been known to mislead night travellers with dancing lights of will o' the wisp. It was not unknown for a man to carry a needle beneath his lapel to ward off the night imps. 'Of course I don't believe in fairies,' a local told me, 'even if they exist.'

Myths seep from this land like marsh damp. The Celts from the east brought child sacrifice in the hope of fair weather and fertility. They worshipped three-faced gods with antlers, and snakes with the head of a ram. Great one-eyed sun gods flared, then faded, leaving only the faint vapours of fairies and hobgoblins. But even they could not be dispersed by Christianity, only bottled and sold as incense; Saint Brigit became one with the much more ancient Brighid the fire goddess, guardian of the hearth.

Until the dawning of the 20th century, a few farmers still threw flaming brands into the fields on May Day to ensure fertility, and drove their cattle between two fires to protect them from disease. If a girl leaped over the flames, she was sure to marry and bear many children. And usually she did.

Saint Ina's well, I was told by the locals, could cure a variety of ills. All that was required was to leave a memento and drop stones into the water accompanied by Hail Marys. This had to be done for three consecutive years, by which time you were cured, or had died in the interim.

A battalion of bats flapped around our ears as we walked home in the gathering darkness, the lough's surface metalled by moonlight. Under the tunnel of trees it was blind black and we held hands so that nobody strayed too close to the steep drop down to the water. Then suddenly the wind got up, the trees shuddered and the forest came to life around us. A long-eared owl swooped in a whisper of wings and a surprised mouse squealed before being swallowed whole.

The water from Saint Ina's well was said not to boil; indeed, if you had the audacity to put it to the test, even the kettle in which you heated it would never again boil *any* water. Of course, we brought some back to camp, but it must have been a dud lot, for it boiled almost before I lit the gas.

*

After a couple of weeks in camp, some students tired of hills and holy wells and longed for the outside world and the brightish lights of Skibbereen. In the early days a party went to see the Strolling Players at the Coliseum and got the giggles half way through *East Lynne*. 'I am always polite to women,' declared the leading man. 'Why, my own mother was a woman.'

But now our highest aspirations to culture could be found in Donelan's pub, so very occasionally we would bunk off into town. Jack, of course, had no desire to accompany us and John also stayed behind, perhaps out of solidarity.

Donelan's was barely a pub, primarily it was the office for 'Ernest Donelan's Motor Garage and Undertaking Establishment'. Ernie's wife, Mary, served behind the bar.

When we piled in on my birthday that year, the bar was crowded and there were warm smiles from the locals. An ancient man and an angel-faced boy of six were playing cards. 'A flush!' cried the child triumphantly, scooping up the money. He then riffled the cards with the confidence of Bret Maverick and dealt a

fresh hand. The old man resignedly examined his remaining change.

In the corner sat an old lady in a traditional black shawl. 'Can I have a glass of water?' she asked in a tiny voice. 'And while you're there would you put a whiskey in it?' At her feet was a dog that had lapped up a pint of porter and was now snoring it off.

The other customers were as unlikely a crew as Skibb could muster. On my right was a short chatty fellow with rubbery ears and teeth as yellow as the keys of a pub piano. To my left was his straight man, tall and thin with immense features. I had cycled shorter promontories than his nose, and on better surfaces. He spoke slowly, as if conserving energy to sustain his nose. He proffered a pack of Sweet Afton filter tips.

'No thanks,' said the short man. 'I prefer cigarettes.'

A third fellow had a tiny nose and no teeth. He was two-thirds tipsy and one-third smashed, and his diction had begun to purée.

'Are ye a fillum star?' he said.

'No,' I replied modestly, 'but I can see why you asked.'

'But ish it strange ye are?'

'Well, I try to be.'

'No, no,' he said shaking his head. 'Ish ye a visitor?'

'Yes, I'm a stranger in town.'

'All the way from America?'

'From England.'

'Ye'd be knowing then all them Irish army uniforms ish made in England.'

'Really?'

'Rilly, but 'tis all right. All them Union Jacks come from Ireland.'

'They say,' said the tall man, 'that for the fiftieth anniversary of the Easter Rising they'll be enlarging the General Post Office in Dublin to accommodate all the heroes who claim to have fought in it.'

The pub made no concessions to decor, comfort or music. The banter was everything. From a counter polished by an eternity of elbows protruded two immense pumps like truncheons. 'What's in that one?' I asked.

'Guinness,' said Mrs Donelan.

'And the other one?'

'That's Guinness too.'

'Then a Guinness it is.'

The pump foamed at the mouth. A good pint of porter needs a thick, creamy head. The pump delivered a great head but omitted to insert the body beneath. Mrs Donelan scooped off the excess froth with graceful sweeps of a wooden spoon, then refilled and rescooped until eventually the pint was born, but it was a difficult delivery.

'Do you know Murphy's stout?' the tall man asked.

''Tis his own fault for eating so much,' said his companion.

'When his wife produced her first, Congratulations, says I to Murphy. Who do you suspect?'

'It arrived nine months to the day. The only contract Murphy ever delivered on time.'

'Maureen was wasted on himself. Wasn't she the most beautiful girl west of Kinsale?'

'I'd say she was.'

'I myself went out with her years ago,' boasted the little man. 'She wore layers and layers of petticoats. 'Twas like playing pass the parcel.'

'Lost her virginity every night, they say, but always found it again by morning.'

Someone rang to arrange a funeral. Mrs Donelan took down the details, announcing them in a loud voice.

'I buried my husband twelve years ago,' said the old lady in the corner. 'He was dead, you see.'

The small man leaned towards me and whispered. 'The black lamb of the family,' he confided. 'He caught the temperance and went all respectable. It was terrible sad to see him refuse a drink – but he recovered in time and died a waster.'

'I was very lucky,' said the old lady. 'I met my husband up a ladder.'

'Less have a sing-song,' bawled the toothless man. 'I've had a grand day altogether. A day I'll nefer forget.'

'And a night you'll have difficulty remembering, I'm tinkin'. You're drinking a ton tonight, Tim. Why would you be doin' that?'

'Well, I can't rely on anybody elsh to do it for me.'

He then began to sing:

Oh nefer, oh nefer, oh nefer again,
If I live to a hunthred or a hunthred and ten,
For I fell to the ground an' I couldn't get up
Ahfter drinkin' a quart 'o the Johnny Jump up.

Says a Gard testin' me, say these words if ye can:
'Around the ragged rock the rugged rascal ran.'
Tell dem I'm not crazy, tell dem I'm not mad.
'Twas only a sip o' the bottle I had.

After the tumultuous applause had died down someone asked, 'What time is the first bus to Cork in the morning?'

The toothless troubadour scrutinised, with difficulty, a

schedule on the wall. 'Accordin' to the skidoodle 'tis eight o'clock.'

'I hated going to Cork and breathing all those fumes,' said the tall man. 'But now we've the traffic here for those too ill to travel.'

'I took my driving test on Tuesday and passed with flying—'

'Pedestrians?' said the tall man.

Earlier that evening a birthday cake had appeared with fifty candles on it – because the cook said she didn't know how old I was and had to guess from my appearance. Now I bought everybody in the bar a drink.

'What will you be having?'

'Too much, I'm tinking,' said the tall man. 'But I'll have a Guinness.'

'May all your sons be priests,' said toothless Tim. He raised his glass and shouted, 'To temperance!' – a signal for everyone to down their drinks.

Mr Donelan came in. 'Will you have one?' I asked.

'No, thank you. If it was drinking I was, I'd soon be my best customer.'

In the old days, I was told, Ernie had converted his old taxi into a hearse by cutting a hole in the back for the coffins to pass through. But now he could afford the real thing. 'How's the new hearse, Ernie?' someone asked.

'Ah it's grand,' said he. 'Soon all Skibbereen will be dyin' for a ride in it.'

It was from Donelan's bar that I watched Apollo 11 land on the moon. The rest of the world was entranced as Neil Armstrong hopped around then returned to the *Eagle* lander. 'The hatch is closed and latched,' Buzz Aldrin confirmed, and Michael Collins orbiting above shouted, 'Hallelujah!' – at least so I'm told. At the time they were broadcasting an interview with Collins' old aunt in Wexford. Whatever celebrity commanded the news, you could rely on them having Irish relatives awaiting interview. It came as no surprise to discover that Che Guevara's granny was a Lynch, and Muhammad Ali's grandfather an O'Grady from Clare.

Nobody was sure what time was closing time, but it wasn't like the old days when the law allowed a traveller to be served whatever the time. He had to come three miles to be a *bona fide* traveller, so after closing time the lanes were cluttered with *bona fides* on tipsy bicycles, criss-crossing the countryside en route to each other's pubs. Not everyone was so conscientious, and the pubs were often open all hours, full of tipplers who had journeyed almost half the length of the street. The *Garda* were very understanding, and it was not unknown for the publican to be told the time of an impending raid so the customers could be temporarily evacuated into the back lane to avoid an embarrassing prosecution. Although those days were gone, it was very late before we left.

It was a long walk home along dark roads with a reckless wind in the trees, so we were relieved when a car stopped to give us a lift. The three students crammed into the back and I got in the front. Our suspicions were aroused when the driver ground the cogs from the gears and the car kangarooed into action. He was roaring drunk and proud of it.

'Me name's Cornelius and I'm stewed,' he boasted. 'I told my wife I was taking the car to the pub. Well, if you're going to drink and drive, says she, you'll be needing the car.'

He could hardly see, but it didn't matter, for he was forever turning round to talk to us. 'I'm a plasterer by trade and plastered for a hobby.'

'Can I help?' I offered.

'The more the merrier.'

So he worked the accelerator and brake while, from the passenger seat, I looked after the clutch and gears and helped with the steering. Thus we swerved along at speed into the night with moths flaring in the headlights' glare like the ghosts we were all about to become. On one side was the mountain, on the other a sheer drop into the cold, inky lough.

'This will do fine, just fine,' I said when we reached the postbox beside the Goleen. Then the plasterer revealed that he lived not on this road, but on the other side of the lough. He would have to turn round. So I climbed back into the car and together we did a perfect twelve-point turn.

'Are you sure you know the way home?' I asked.

'Ach, there's no need,' he said. 'The car's been there before.'

Kerry

In spite of the good times, friction was often lurking just beneath the surface. One evening some of the lads caught a dozen mackerel – a rare event. Over dinner Jack decided we should have them for breakfast the next day.

'That's not on, Jack,' John argued. 'We are going to Kerry tomorrow, leaving at six-thirty in the morning.'

'That's no reason,' Jack insisted. 'I shall get up early and cook them.'

'Jack, it will take far too long and then there's the washing up – greasy pans and plates. The taxis won't wait. We can't possibly do it tomorrow.'

'Of course we can, no problem at all.'

Exasperated, John leaped to his feet. 'Well, I'll have nothing to do with it,' he declared and strode out.

John's wife, Erika, our cook, got up in sympathy and said,

'Oh, Jack,' throwing open her arms in a gesture of despair, forgetting she was holding a cup. The coffee leapt across the room to paint a brown exclamation mark on the door. Then she too ran out.

There was a stunned silence until, to everyone's relief, I said, 'Shall we wash up?'

To my surprise we went up to the Bohanes' house that evening as if nothing had happened. But something *had* happened; Jack had been reprimanded in front of his students. To the best of my knowledge the incident was never mentioned again, but something unspoken is not something forgotten.

The next day, after a wash in icy water, I felt as alert as I was ever likely to be at six in the morning. A light breakfast fortified us for the military-style expedition that was a day's outing to Kerry. We were ferried in batches to the postbox quay, summoned by a barely audible car horn. 'Loud as a mouse's fart,' John observed. 'Ernie's let the battery go flat again.'

Four taxis awaited us. John always went in Ernie Donelan's car, but Jack, less brave, preferred to be driven by Ernie's daughter, Pat, or his son, Kevin. Ernie's car led the way. He signalled to pull out and the light winked for the next couple of miles.

We drove north to Kenmare, then on through rough-and-tumble terrain up the steep winding road into the hills. This land had been squeezed and crumpled into long sandstone ridges, running from the mountain spines in the east to long promontories in the west that reached thirty miles into the Atlantic, making the coastline twelve times longer than it need have been.

The hills were once clawed by ice, and the rivers had been

turned at right angles and sent to search for the sea sixty miles to the east. Glaciers had lingered on the slopes, scooping out great basins of rock. In one of these corries not far from the road was a cold black lake embraced by an armchair of cliffs.

We broke through the arête at windswept Moll's Gap, and turned along the ridge to picnic at Ladies' View beside a rock that proclaimed in large letters: AFTER DEATH – JUDGEMENT.

Far below us lay the lakes of Killarney in their great valley, one basin scooped by ice, the other dissolved by rain. The large lower lake is dotted with islands, once the fortress homes of chieftains and the retreats of monks. On one is the monastery where the monks educated Brian Boru, who became the first king of Ireland and drove out the Vikings at the cost of his life. On the lake shore Saint Brendan the Navigator was born in 489AD. He is said to have crossed the Atlantic in a hide-skin boat, but, having seen what sea water did to my leather shoes, I doubted it. In 1813 Shelley spent three weeks in a cottage on one of the islands and warned of the 'perilous navigation of the lakes . . . of sudden gusts', and how 'vessels were swamped and sunk in a moment'. What a pity he failed to remember this before his final sea voyage only nine years later.

Tourists were ferried around on jaunting carts at five miles per hour, under the impression that this was how the locals got to the shops. The carts carried 'four if ye sits dignified and six if ye sits familiar'. In earlier days it was commonplace for a passenger to fall off. Indeed, Thackeray claimed the carts were *designed* to fall from. He himself came adrift together with a very pretty woman who 'showed a pair of never-mind-what-coloured garters', and concluded that, 'considering the circumstances of the case, and in the same company, I would rather fall off than not'.

I once heard a cart's jarvey (driver) plying ageing Americans

with tales of the rejuvenating qualities of the waters: 'Why, one old buffer fell in and was no more than eighteen years old when he came up.' He was so convincing it's a wonder they didn't dive straight in and drown on the spot.

But all the blarney in the world cannot spoil the beauty of the place.

The lake-reflected hills were once as high as the Himalayas. Now relegated by erosion to a manageable magnificence, they were still the tallest and wettest mountains in Ireland. The McGillycuddy Reeks is a ludicrous name for this wonderful jumble of peaks dissected by sombre valleys. The lower outcrops had been ice-smoothed into the shape of hunched animals, but the highest peaks had protruded above the ancient ice sheet and were shattered by frost into pinnacles of slate.

Jack would walk over Purple Mountain as he always did. Equipped with a compass and large ginger boots on the ends of his spidery legs he strode off, accompanied by two adventurous students. At first they tramped through sparse woodland carpeted with mounds of moss – three quarters of the entire bryophyte flora of the British Isles can be found here. Beyond, they came to a lime-green meadow of moss spiked with dark sedges and dimpled with soggy hollows in which carnivorous butterworts averted their violet faces from the carnage in the leaves below. Even in the sponge that is Ireland, this region is famous for marshes. Sometimes bogs, naturally dammed by peat, burst out, and an avalanche of slurry swilled down into the valley – a peculiarly Irish phenomenon and potentially lethal. In 1896 a farmer collecting fuel dug too deep into a peat dam, and he and his family of seven were swept away with all their animals. The huge bogslide did not stop until it emptied into the lower lake of Killarney. The family were never found and still lie somewhere, entombed in the peat and perfectly preserved by the acidity of the soil.

Jack's small party strode up the steep rise of the hill past streams shallow with summer, but containing just enough water to glide the sun. Here was the haunt of those relict plants that had escaped the ice: great saxifrages, the Irish spurge, and others whose nearest relatives were in Portugal and North America. But on top of Purple Mountain there was only huddled heather and cotton grass shivering in the wind. The summit was a rock-strewn wilderness, the abandoned yard of a monumental mason. It was these fields of dark scree that gave the mountain its purple head. There was no path, no cairn at the summit and, like all the high hills of Kerry, it was bereft of people. They were the best kept secret in mountaineering.

'Watch out,' Jack warned. 'If the rock looks slippery, it is. If it doesn't, it probably is.'

Up here there was just the whistle of the wind and the croak of a raven. For a while they sat on the sun-warmed slates and admired the sharp, shining landscape. But these peaks are forever lost and found in the mist, and Jack noticed the dark amoeba of a storm cloud looming towards them.

They hurried on as fast as they dared towards Tomies Mountain, with its cluster of Neolithic tombs on top and flanks of purple sandstone, but they were soaked long before they reached the deep gash of the Gap of Dunloe.

The rest of the party were driven towards Killarney but had shunned the most touristy town in Ireland, where everything on offer was green or shamrock-shaped or both. It had thousands of visitors each year, but nowhere to park. Instead we went to the Gap of Dunloe – landscape become cliché.

The Gap was formed as the ice melted and a torrent of water flushed away the land, leaving a chasm flanked with cliffs. It had long been a tourist attraction. Shelley came, as did Charlotte Brontë on her honeymoon, the passionate author still calling her husband 'Mister Nicholls'. She fell under her pony and 'felt her kick, plunge, trample round me . . . It was mere good fortune I was not trampled to death'.

The ponies remain. Sean O'Faolain warned how 'a flock of banditti' fell upon him 'out of a whirl of coat tails and black hats, shaggy moustaches and waving ash plants,' to entice him on to ponies whose 'saddles were equally worn and wearing'. There had always been bandits here to waylay the unwary tourist. In the 19th century Thomas Cook tried to circumvent them by taking coaches up the Gap, but they blocked the road and shot at any carriages that got through.

Hustlers disguised as ostlers jostled for our custom. They were born with caps on, and in hand.

'Is it a horseman ye are, sir?'

My bowed legs had misled him.

'Then this is the steed fer you. Lightning's his name an' lightning's his nature.'

John had trouble mounting and when he succeeded he missed the saddle. But soon we were all aloft like sacks of peat on an assortment of moth-eaten ponies. They were of indeterminate age as many of their teeth were missing.

I was handed a switch. 'If he don't go then beat 'im,' was my only driving instruction.

So Lightning was struck twice in the same place and ambled forward at a resigned dawdle. I dug in my heels and so did he. He tilted his head and eyed me and I knew he had broken stouter hearts than mine. I leaned forward and whispered in his ear, 'If you don't do as you're told I'll sell you for horseradish sauce.' Having only limited botanical knowledge this put the wind up him and from then on he was fine.

Steering was unnecessary as the ponies had been up the Gap a thousand times before. Most of them walked all the way except when they stopped to browse. Even so, the trip wasn't uneventful for all the posse. The girth strap on Sue's nag came loose and she slumped to starboard, that on John Ebling's snapped and he descended to port, Louise's horse lurched to the left but she went straight on to pirouette gracefully on the ground on her back.

The track winds for several miles, steepening all the time. We passed a litter of glacial debris, huge boulders and erratics, and skirted the shores of a dark lake fretted with water hawthorn. Gradually the cliffs closed in on either side and the slot of sky above bruised to indigo. Then the rain and lightning came. The heavens cleared their throat and great wet gobs clattered down. Every fissure in the cliffs above us spouted water and cascades appeared everywhere.

Sudden downpours are commonplace in the Reeks, and can raise the rivers three metres in as many minutes. Within moments we were soaked. It was torrential, it was magnificent, then it was over and the sun streamed back into the gorge. My dead-beat horse now smelled like a dead horse, and steamed like a compost heap on stilts. As the grade got steeper, Lightning slowed. I could have walked faster and I knew damned well *he* could.

At the top of the Gap there was a wonderful sweep of a valley and the gleam of distant lakes. I reluctantly turned my nag and jabbed his flanks with my heels and he was away. It was a bolt of Lightning. He clattered down the steep slope slipping on the wet stones regardless of his safety or mine. We scattered hikers and caused horses hauling gigs to rear and spill their charges. Lightning was heading home and he had no brakes.

Back at base, Lightning clattered to a halt in a minor dust storm and even the banditti were impressed. I was indeed a horseman. I sat tall in the saddle for a moment, then, cool as Clint Eastwood, dismounted and unsteadily regained my feet.

We assembled at Kate Kearney's Cottage, a touristy pub at

the mouth of the Gap, where Jack's walking party joined us. All bottoms were sore, two pairs of jeans had split and one girl who had ridden in shorts had inner thighs like raw liver, but a couple of Irish coffees magically anaesthetised all our ills.

Lough Iners had been doing this trip for years. In 1950 Jock Sloane had recorded his impressions:

> Some walked, while others rode with zest,
> And two, more crazy than the rest,
> Set off to reach the mountain top.
> They climbed and climbed 'till they could drop
> Of fatigue, and neither knew
> Who'd have the strength to carry who.

But John was more direct:

> One of the girls had a go
> To ride up the Gap of Dunloe.
> But she had chosen a pony,
> Whose back was so bony,
> That she wore out her trousers below.
>
> A couple of curious twits
> Kept the whole party in fits.
> While we laughed like drains,
> They lost hold of the reins,
> And had to hang on by their wits.

Ernie Donelan had ridden the Gap one year, but never again, for he hadn't been able to sit down for a week.

We decamped to Glengarriff, thirty miles to the south, in

cars heady with the aroma of ripening stables. Glengarriff lies on the shores of one of the most beautiful bays in Ireland. People were eating oysters here 3,000 years ago, and left a shell midden thirty-three metres long, three metres wide and almost a metre deep.

We slipped into the Eccles Hotel, where Shaw is said to have written *Saint Joan*. In the washrooms we stripped off and tried to dispel the smells of the day. The girls were mistaken for a travelling cabaret.

We had entered as caterpillars, and emerged as crumpled butterflies. Our best clothes had been bundled up inside backpacks for a fortnight, and the concertina pleats saw no reason to fall out simply to save our embarrassment. Jack looked smarter than I had ever seen him, although his tweed jacket was a trifle too small and his trousers too short and his tie, in the colours of a pedestrian flattened by a steamroller, went with nothing he (or anyone else) was wearing.

The bar was filled with snooty types swilling Paddy's whiskey and dressed to death except for their little yachting wellies. Their nautical caps bore impressive badges somewhere between the Submarine Service and the St John Ambulance Brigade. The lounge bulged with a coachload of American dowagers. 'Blue-rinsed biddies who have driven their husbands into early graves and are now touring Europe on the proceeds,' John announced.

'I can hear you, young man,' croaked an ancient American voice from a dim recess.

The Bantry Bay salmon was excellent and so were the strawberries and fresh cream. The entire three-course meal cost about £1, and John treated us to bottles of Matéus Rosé.

Across Bantry Bay we watched the sunlight fade on Whiddy

Island and the violet-shadowed mountains beyond. Gulf Oil were building a terminal on the Island, but it would end spectacularly when the tanker *Betelgeuse* went supernova and took the terminal and a corner of the island with it. The blast would shake Skibbereen, far away over the mountains.

We didn't leave until long after the moon had risen and, exhausted, we slept all the way back. When we entered the darkened laboratory we were welcomed home by bottled plankton samples glowing eerily green in their jars.

Projects

The lough is ideal for field work. Whichever way the wind blows there is always a sheltered shore in the lee of the summer snowstorm of thistledown, and calm waters enabled us to operate from small boats. As the ancient paths became overgrown and impassable we were increasingly isolated, and boats became the only means of transport to the sites.

Rowing was one of Jack's pleasures at Lough Ine, and he had unparalleled ingenuity for devising excuses for picnics at sea, whatever the weather. Sometimes you heard the rain coming, beating over the trees with the clatter of hooves, then advancing over the water towards you. I remember bullets of rain punching holes in the surface of the sea and Jack merely donning his sou'wester and continuing to munch through his sogging sandwich. 'Looks like rain,' I said.

Only rowing boats were allowed because they could work in

the shallows, could be lifted out of the water at the end of the season for winter storage, and they were quiet – except, of course, when Jack was at the oars. The brown boat had developed his style. The best corrective for its propensity to pirouette was a short chopping stroke that rattled the rowlocks: *topocher ker plunk, topocher ker plunk* . . . This contrasted with the call of the two sturdy black boats. Instead of rowlocks they had thole pins – long, paired dowels that stuck up from the gunwale. When heaved, the leather-sheathed oars rose between the pins, then fell heavily on the backstroke to club the gunwale: *thud didi thud, thud didi thud* . . . Later, Jack bought a fibreglass boat, so flimsy that its floor rippled violently when underway as if there was nothing much between you and the deep. It was a versatile craft – no use at all for almost everything.

Most students learned to row and Jack took pleasure imparting the intricacies of the clove hitch for tying the painter to a bollard, and the rolling hitch – 'the only knot that should be pulled tight' – for tethering it to the out-haul.

The locals puzzled over our activities. One confidently informed a visitor that we stuck unfortunate cockroachs on pins to see if they had two livers. Most hadn't a clue what we were doing . . . and sometimes neither did I.

The field work was carried out in summer instalments with several projects often running simultaneously. Occasionally one got lost. In 1965 we had saturated the bottom of Barloge Creek with crab traps, capturing hundreds of crabs, then dabbing them with paint and returning them to the exact spots from whence they came, only to attempt to recapture them later. We re-potted the area repeatedly, the idea being that the proportion of marked to unmarked individuals refound would enable us to calculate the size of the total population, and the position of recapture would indicate how far the animals moved and mixed with each other. The effort of transporting boatloads of traps down the Rapids, then baiting, laying and retrieving them was immense, but nothing ever came of it.

Some projects, however, were models of logical investigation – like our study of the jewel anemone, *Corynactis*. If you dive in dark places or look beneath boulders, you find these jewels twinkling like stars in the current. They split repeatedly to form patches, each with different bright decor: lime green with pink tentacles, magenta, orange, all with contrasting trim and tentacles with pearl-white knobs. They are among the most beautiful creatures in the sea.

But why did they hide away their loveliness in the dark? Why not on the tops of boulders where their colours would glow?

We transported them, in submerged panniers on either side of the boat, to sunny sites and followed their fate. They fared badly, but was it because of excessive light or the accumulation of silt?

We then put *Corynactis*-covered boulders on spiked 'Gandhi beds' to keep them out of the silt. The rocks were placed anemones-up or anemones-down to vary the light regime. In 1968 we constructed the 'battleship', a floating stretcher which held large, black plastic boxes containing anemonied boulders, some shaded with lids. Every thirty minutes for a day and a night we rowed out and counted the number of anemones that were open or shut. One student dropped a lid and it sank from sight. I had instantly to dive over the 'battleship' to retrieve it before it vanished into the depths for ever.

The following year I also removed a large area of over-shadowing kelp from the Rapids and dived there each day to monitor the anemone populations. Soon a lawn of algae sprang up. Its fluttering in the current continually irritated the anemones and kept them shut, just as for us sleep would be impossible if a fluttering curtain continually blew across our face. Slowly, they shrank away and died.

In the laboratory *Corynactis* could be persuaded to open and close to order simply by altering the light, and with colour filters we determined which wavelengths of light did the trick. By this time we had 'corynactivated' to exhaustion, but had unravelled much of the mystery of the anemones' behaviour and ecology. This may sound a small matter in the grand scheme of things, but at this very moment in all the world's tropical seas, uncountable numbers of tiny polyps, the builders of great coral reefs, are opening as dusk falls or closing with the dawn, just as *Corynactis* does.

This was typical of the thoroughness of our investigations. Problems were attacked from a dozen different angles until they all pointed in the same direction. Observing Jack at work taught me how to do field research. He had not just the passion – the academic's need to know – but also the practical nous to devise ways to find out, and the energy to do it no matter how long it took. Not many mental sprinters can also run the marathon.

The Western Trough is deep and calm, and in summer the waters from the ocean glide in, bringing a litter of edgy plankton that flits about on the invisible barrier separating the warmer waters above from the cold depths. Like the Rapids and Bullock Island cave, it was a 'natural experiment', one in which we could monitor the seasonal development of the stratification and the consequences for the bottom-dwellers of the summer fall in oxygen.

We placed temperature probes at ten different distances from the shore, the nearest ones in the shallows, the distant ones in the depths of the trough. The cables leading to the probes would be vulnerable underwater, so Jack determined they should be encased in plastic tubing, three-metre lengths clamped together to form a continuous pipeline. He ordered 1,700 metres of alkathene tubing. As the longest of the ten cables was over 200 metres, threading the cable was a huge task. Relays of students threaded for hours until their fingers were threadbare.

Then the cables had to be placed in the lough. One at a time, the piped lines were coiled into a boat in a towering black plastic doughnut two metres high. The cable was slowly paid out as the boat was rowed offshore, and was anchored at intervals to concrete blocks lowered to the sea floor. Finally, the probe at the cable's end could be fixed at the appropriate depth.

Through the summer of 1969 the trough was monitored night and day, like a patient in intensive care. In the attic of the mess hut, ranks of machines brought from Norwich would automatically record the temperature at each depth every five minutes. But they rarely did. Their batteries conked out or the paper jammed, or the leads fell out, or damp seeped in. Recorders that would operate unattended for months back in the university laboratory took a dislike to working abroad. Jack brought over Peter LeFevre, an electronics technician, and although he laboured in the dim attic until his face became gaunt and his eyes bulged, as soon as he wasn't looking the machines went on strike again. Eventually we resorted to taking the readings manually. Shifts of students manned the attic recorders day and night at critical times of the year.

In addition, we rowed out to marked points in the lough and took temperature, salinity and oxygen readings at one-metre depth intervals from the surface to the bottom.

'Have you two been probing recently?' Jack asked a couple

who were holding hands. Their faces said they had.

'Why, no,' they replied, too emphatically.

'Then do a profeel at Buoy B.' Jack always said 'profeel' for 'profile'.

'Certainly, Jack,' they said with relief.

'Make sure you joggle the probe or it won't work properly,' Jack shouted after them. I had to look away, a snigger would have landed me in big trouble.

'And don't forget to take a sheave with you,' Jack added.

For a moment they thought he said 'a sheath', until they remembered that the cable had to be let down over a measuring reel to make sure the probe went to the correct depth.

They rowed out together for a shipboard romance and found a dead goat floating at Buoy B.

With relays of TWITS – Temperature Waves in the Thermocline Surveyors – we were able to pinpoint the days when the stratification formed in summer and broke down in the autumn. The fate of the bottom fauna was monitored by sampling with a grab, a jawed bucket that takes bites out of the sea floor. Every sample had to be passed through a series of sieves to determine the proportions of mud, sand and gravel. These were oven-dried and weighed, but not until every animal had been extracted. We identified 112 species from ninety-six samples; one sample alone contained over 600 worms and 5,000 tiny snails. It was a big job and was supplemented with experiments on the survival of animals above and within the deoxygenated zone.

Millions of dollars were being spent on sending rockets out into the dusty wastes of space to fall far short of the Milky Way. Jack found as much magic squirming in the muds of Lough Ine, and did it on a shoestring. Although he generously invested in the laboratories, what he spent most freely was ingenuity and time. I came to understand that ecological studies must keep in pace with the rhythm of the system: the urgency or lethargy of colonisation and growth, and the time it takes for living creatures to resolve their differences and exert their influence.

Projects carried out over only a few months each summer took a long time to complete. The *Corynactis* study began in 1955, but seventeen years elapsed before it was written up for publication. Everything militated against bringing the work to a rapid conclusion. Usually several projects ran simultaneously and vied for our attention. The study of the Western Trough began in the same year as a project on the ecology of the purple sea urchin. One took eleven years to complete, the other eighteen.

Some tasks had to be repeated, because we suspected the students hadn't done them properly. This was not surprising, as every year there was a new batch of first years who were well versed in biochemistry and genetics, but knew little of animals and plants or research methods. Clearly, more senior students would be better suited to our purpose, but for some reason it was mostly freshers who came.

Sometimes Jack shelved a project only to take it up again years later. For example, underwater surveys of Whirlpool Cliff were carried out in several summers from 1967 to 1972, but nothing further would be done until 1982. It would be twenty-three years from when the first samples were taken until the paper appeared in print.

In those days the phrase 'publish or perish' was fresh and menacing. I was a young lecturer whose promotion at Glasgow University depended chiefly on the quantity and quality of my publications. But urgency rarely infused the work at Lough Ine.

My solution from 1970 onwards was to skip some of the excursions and, instead, to set up modest experiments of my own

– things that could be tended in spare moments or in the evenings, and completed within a few weeks. I looked at red algae that lost their regimented growth form in the darkness of Bullock cave, and later, in the laboratory in Glasgow, I helped them to re-find it by providing appropriate light. I sought to discover why my old friend *Saccorhiza*, the kelp that dominated the Rapids, did not colonise the lough. Another project elucidated why a colonial organism living on kelp grew predominantly downwards on the plant. Later, an expert reviewing what was known of these animals praised the ingenuity of this last project, but Jack dismissed them all as 'quick-offs'. This was the phrase he used for short-term, and therefore superficial, studies, as compared to the marathon team projects at Lough Ine.

Underwater

Although little changed at Lough Ine, by the early 1970s Northern Ireland was in turmoil. The IRA had once affectionately been known as 'I Ran Away', but no longer. Every Left Luggage office in England was closed in case someone decided to deposit a bomb. But not in Southern Ireland. Even so, when I tried to deposit my backpack at the Dublin railway station, the attendant eyed me suspiciously. 'What's in here?' he snapped.

'Just personal belongings,' I said, 'clothes and things.' I didn't mention my huge diving knife.

'Just clothes,' he said with obvious relief. 'Thank God for that.'

At the Guinness brewery, I eventually got a free drink of stout, and in my euphoria almost forgot to reclaim my luggage. 'No, not that one,' I said to the attendant. 'The grey one with the Semtex in it.'

✳

I never got to see much of Cork City, but I fitted in an instructive visit to the bus station lavatory and read the lively debate on the door:

Release IRA prisoners who are rotting in prison.
And may they all rot in hell.
That's all we need – a Brit lover in Cork.
All who live by the sword will die by the sword . . .
    sometimes anyway.
You would think the Provos would see the Eire of their ways.
The Brits are dying on Irish soil.
But they are keeping the peace.
Aye, keeping the pieces of dead Catholics.

The last correspondent attempted to sum up:

By the way you write, Satan was here last night.

The Troubles had impinged little on our work in the south, but then in the winter of 1971 there was an incident. A Dutchman who lived over by the coast not far from the lough foolishly closed all the traditional rights of way across his land. One night someone accidentally ignited his house and, it was said, inadvertantly fired at him when he tried to flee. The fire engine had to drive across the beach to get to the house, but was confronted by a sign: STRAND MINED BY THE IRA.

It wasn't, of course. But, just in case, the brigade diligently rolled slow stones in front of the appliance all the way, and by the time they arrived, the house had burnt to the ground.

It made Jack worry about bringing students into possible

danger. So in 1972 he took them instead to Loch Torridon, a stunning site on the west coast of Scotland, and later in the summer just he and I and a group of young research divers from Oxford and Cambridge went to Lough Ine. It was strange without the students and John, and it brought home to me how much they contributed to the fun of our summer expeditions. None the less, although there were only six of us, we laboured night and day and did twice as much useful work as in a normal year, when thirty or so students came too.

In theory, collecting samples, even underwater, is easy, but sorting and identifying them is difficult and time-consuming. So, with five or six divers bringing up material we should have been swamped with work, but rarely were. Divers spend very little time in the water. Mostly they are occupied patching torn diving suits or rubber boats and sealing leaking camera cases. They spend long hours replacing 'O' rings and mending machinery, for bits fall off and the air compressor invariably has a vital component mysteriously missing and unavailable in Ireland.

You can always detect a diver in the dark; he reeks of Evostick and oil . . . and beer, for all these frustrations drive him to drink.

When you dive too deep the nitrogen that dissolves in your blood intoxicates. At forty metres you feel light-headed, at fifty you become dizzy and careless; below sixty you're in trouble for your judgement is impaired and you may offer your mouthpiece to a fish under the impression that it's the one who's drowning. It is called nitrogen narcosis, the 'rapture of the deep'. Surprising, then, that divers also need the oblivion of alcohol – the rapture of the shallow – yet in any seaside pub there is an Evosticky corner where they congregate to swap their experiences with glues, compare their missing parts and tell astounding underwater tales

– some of which may even be true.

Jack still felt that the rhythmic rattle of oars was the maximum disturbance the lough should suffer, but the divers, when they worked out at sea, were allowed to use a motor. So they brought an outboard as big as a Labrador's kennel. When they opened the throttle it roared, and the Zodiac inflatable reared like a rodeo stallion as it galloped away at speed. Well, sometimes it did and sometimes it didn't – for as often as not there was an air bubble in the fuel line, or a nut had escaped into the works, or the shear pin had shorn. Those shear pins were excellent; they would shear at the slightest thing. When it came to shearing, they couldn't be faulted.

Starting was a different matter. It started when it felt like it, not to suit us. Threats and swearing didn't help. Crying rarely worked, but if you could convince it that you were not just angry, but in such despair that all that was left was to fling it into the deepest pit of the ocean and drown it for ever, then and only then, would it leap into life and chuckle to itself as if to say, 'Come on, what are we waiting for?'

The locals didn't associate engines with marine biologists and thought the divers were after their lobsters. One day, as the divers' boat eased out along Barloge, shots whizzed overhead and they had to throttle away out of range.

John Gamble, one of the divers, had a tape recorder in a sometimes waterproof box. He tended its seams every day with aquarium sealant. The idea was that scrawling on a plastic pad would soon be a thing of the past; instead, he would simply dictate his notes on the ocean floor. One evening I saw him huddled over the machine, trying to transcribe a tape. As far as I could tell, it had recorded the hoarse raspings of a heavy breather against a background of someone farting in the bath. There were also intermittent mumblings from a drunk in an echo chamber. He squinted at the machine, his brow concertinaed with concentration, as if trying to decipher transmissions from Mars. He might as well have been.

The divers returned every summer for the next few years, and to investigate the range of deep underwater habitats available in and around the lough we dived at three very different sites: Whirlpool Cliff, Carrigathorna and the Western Trough. Whirlpool Cliff stands inside the lough opposite the Rapids. During tidal inflow, a jet of water streams in, is deflected by the cliff, and forms an anticlockwise gyre. It is not likely to suck unwary ocean liners to their deaths, but it took our boats and lazy jellyfish for a spin at almost a metre per second.

We knew the track of the water exactly, because in 1954 a boat had been tethered across the Rapids. It had a cross plank amidships so that a platform projected on either side. At each end of the plank were large tubs of harmless green dye. When Jack

blew his whistle, John Ebling rocked the boat (he was good at that) and upset the dye into the water. He accidentally capsized the boat and had to swim for it in a fluorescent green plume that flowed into the lough. The locals, alerted that the 'mad doctors' were to attempt something spectacular, lined the cliff above and cheered.

The dye swirled around the Whirlpool where other boats were moored, each with a spaghetti tangle of rubber tubes suspended at different depths. For three hours water was sucked up with old Air Raid Precautions stirrup pumps so that the concentration of dye at each depth could be determined. It was detectable even when diluted by the sea down to one part in a million. Thus, thanks to Jack's flair, the path of the water and the degree of mixing was accurately plotted – with equipment that every other household in Britain was chucking out now it was no longer required for quenching incendiary bombs.

The cliff is nearly vertical and drops eighteen metres to the shell gravel and rubble floor of the whirlpool. At intervals there are horizontal ledges bearing wonderful gardens of kelp, like watery window boxes on a high-rise block. We measured and weighed the plants and brought up large boulders (labelled 'the Republic') from the whirlpool floor. There was also an unbroken jar that had toppled into the lough from Renouf's wrecked hut twenty-five years before. In warm years we were joined by tiny sea horses and tropical trigger fish whose sneer revealed the rats' teeth they used for cracking open lobsters.

Carrigathorna is a twenty-seven-metre deep underwater bluff that strides out into the Atlantic to face the waves, making diving possible only on the calmest of days. It was a place to see exotic oceanic drifters: seaweed from the Sargasso Sea, the bloated lavender float of the Portuguese man o' war, its sail crimped like a Cornish pasty and with fearsome trailing tentacles below, and a bewildered turtle a long way from its home in the Caribbean.

With two divers on the surface, someone in the boat shouted 'Shark!' and they leapt from the water like frogs from a hot pond. Sure enough, there was a large triangular fin protruding from the waves, but it seemed sort of floppy, too listless to be a maneater on the prowl. When we rowed closer we saw it wasn't a shark, but a sun fish, a harmless vertical dinner plate of a fish, two metres in diameter. Luckily, none of the real sharks we occasionally saw seemed to think of us as something for *their* dinner plate.

We were measuring the amount of light reaching the seaweed at different depths. The kelp cut out over ninety per cent of it, and beneath the canopy there was perpetual twilight. Beautiful, light-hungry carmine seaweeds perched on the top of the kelp stalks seeking the sunshine, but those on the rock below made do with the brief but brilliant flecks of sunlight that flashed down when the canopy parted.

Jack sent me down to retrieve the light-measuring photocells and, while waiting in the boat for me to return, he was accosted by a local, curious as to why someone without a rod or pots should be bobbing about on the ocean.

'Is it fishing you are?' he asked.

'No,' replied Jack at his most precisely obtuse. 'We are monitoring irradiance amidst *Laminaria* in the sublittoral zone.'

'Ah yes, I was thinking that's what it would be.' But the local man circled around close by to see what was *really* going on.

I surfaced and handed Jack the photocell. In a loud voice he asked, 'Is the warhead intact, Carruthers?'

While I pondered what he was talking about, I noticed a chap rowing hell for leather to the shore. Next day there wasn't a person in Skibbereen that hadn't heard of the misguided missile that had plunged into the sea off Lough Ine and made the whole place radioactive.

The Western Trough inside the lough is fifty metres deep, well into the nitrogen narcosis zone, and close to the limits of compressed air diving. In summer, when the sun warms the surface water, the depths remain cold and dark. As I dived down I was shocked by the sudden ten-degree drop in temperature as I crossed the transition zone.

The mud cliffs were impossibly steep, and deep down the water was an eerie luminescent green colour, as if lit for the entrance of the Demon King. And it was absolutely calm. Divers used to being jostled and buffeted by the waves often found the trough's stillness unnerving. For 8,000 years silt had washed in from the surrounding land and caused a relentless fall of dark 'snow'. The bottom sediment was already seven metres thick, some of it so soft that you just sank in as if there were nothing there. As the black cloud enveloped me and closed over my head, I tried not to panic. But which way was up? Which way was out?

Where the silt was firmer, I came across what looked like a sharp-edged bitemark in the unending plain of mud, where we had taken a grab sample the year before.

Nothing much happened down here, except for death. By autumn the bottom-dwellers had used up the oxygen and almost everything had died. Tiny tubular worm coffins bristled from the mud, and abandoned burrows gaped like mouths gasping for air. I was suddenly aware of the air escaping from my demand valve and realised that I was the only living thing in the landscape. So why did I have this feeling that someone – or something – was following me?

> Down! Where the mud is murking,
> Down! Where the worms are lurking
> Silently for me . . .

I was edging towards the brink of panic and was breathing heavily. Above me trembling medusae of air rose, expanding towards the surface, and I became aware that there was more beauty in the rippling rainbows of liberated bubbles than was incarcerated in all the art galleries of the world. A touch of the 'rapture of the deep' imbues great importance to slight things. Even so, long after the rapture had receded, the magical image remained.

Twenty-eight years later the Society for Underwater Technology invited me to tour the country lecturing on the history of science diving. Naturally, Jack's pioneering dives were part of my story. The tour ended at the University of Wales at Bangor, and John Turner, by then a Lough Ine regular, gave the vote of thanks. He presented me with a bottle of wine, a superb book of underwater photographs and . . . the remains of my old French-made diving suit that I had abandoned in the attic at Lough Ine in 1977. He had been using it to patch his suit that summer. I had first taken it to the lough over thirty years before and it had served me well – and still does. As I write this, my computer mouse glides over a neoprene pad that proudly proclaims it was manufactured by:

TARZAN

Marseilles

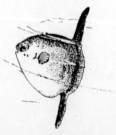

Tensions

Jack was due to retire in 1974, and the future funding of the field trips to Lough Ine was in doubt. His replacement at the University of East Anglia was Tony Dixon, who had been at Glasgow with me. Tony and I met at a party, and he quizzed me about Lough Ine. I enthused, and advised him to go and see for himself, and he did. This was a critical inspection, for if Tony was not impressed, that would be the end of support from the university. So Jack ignored him. He rushed around using his 'I'm busy' jog, too pressed to explain what was going on. Perhaps he was trying to convince Dixon that he was still vigorous and that the field trips weren't just a jaunt to the seaside.

It was left to John to show Dixon around and outline the past achievements and current work. Tony later recalled that the trip 'was made tolerable because of the support and encouragement I received from John Ebling'. Fortunately, in spite of Jack's actions, Dixon agreed to continue the funding.

For Jack and John their ecological work had been a hobby, not the main thrust of their research. Everything changed when Jack retired from his chair: Lough Ine became the focus of his research. For the first time his work there had to be sustained by grants and completed within a set time.

Suddenly, urgency was in the air: 'The specimens must be identified immediately,' Jack demanded. 'This is urgent and is holding up all further progress and planning.'

Now that Jack had to answer to his sponsors, the Royal Society, the authorship of our joint papers became a matter of concern. Being first author on a scientific paper is important; it rightly indicates who did the most work. But if there are more than two authors, not all of them are named when it is referred to by other researchers. A paper by, say, Norton, Kitching & Ebling would be cited as Norton *et al.* (meaning 'and others'), and the 'others' can become relegated to anonymity. In the past, when deciding on the authorship of joint papers, Jack had often erred on the side of generosity, especially towards his younger collaborators, for he knew how important such things are early in your career.

But soon the success of a university would be measured by how much money it processed, and research had already changed from being a game to be enjoyed, to a competition to be won. This urgency had taken a long while to reach Lough Ine, but when inevitably it did, something was lost.

Jack made it clear to John that participation in the field work no longer automatically qualified him for inclusion on all the papers:

'I am contributing far more than any other senior member can . . . I don't consider that you should be concerned as an author in this [project] whether or not you take part in any of the field work. This arrangement would reflect more equitably the contribution made to the overall programme.'

The authorship of particular papers was regulary reviewed. Originally Jack planned to write up our work on the fluctuations of sea urchins and *Codium* with John and me as co-authors. He made it clear that he would be first author, as by far the bulk of the work had been organisation of the student field trips and the identification of the animals and plants, not the field work itself. This was undoubtedly true and I raised no objection, but later it was revised again in a letter to John:

'I had fully intended that you (and Trevor) should participate in the account of the ecology of the urchins, but this was on the assumption that you would take part in the further fieldwork required to make a satisfactory paper on this important and complicated subject . . . In any case . . . probably 99% of the work is done not at Lough Ine but at Norwich by me . . . I plan to write it up and . . . to make acknowledgement of your participation.'

When the paper was eventually published, our formal recognition for contributing twelve summers to the project was merely an acknowledgement that Ebling and Norton took part in some of the earlier visits to Lough Ine, including occasions of the annual census of sea urchins and *Codium*. I was saddened that we had been omitted.

Collaboration is not always easy when, as with the Lough

Ine work, most of the partners have major commitments elsewhere. Jack became increasingly impatient with us: 'I want to get this paper completed during my lifetime. If I don't I shall haunt you.'

It was sometimes difficult to keep up with his demands, as one of the divers found:

> 'I keep getting long scrawled letters from Jack about the Carrigathorna paper. Trying to answer his questions completely and immediately, dropping everything to rewrite something for the umpteenth time — whilst receiving no replies to requests for information — is bad enough, but when he accuses me of not joining into the team work, I feel like packing everything to do with Lough Ine into a tea-chest and sending it to Jack with a note telling him not to bother writing again.'

On this occasion I was able to smooth things over, but the bush was not something Jack used to beat around, especially if he thought the data we had collected was incomplete in some way: 'Am I right in guessing that the reason for giving a standard weight was pure laziness on your part, or am I to blame the divers for not having collected the holdfasts, or probably both?'

John was interested in many topics outside his speciality. 'Thank you very much indeed,' Jack wrote to him, 'for the complimentary copy of your symposium on aggression. It is interesting to think that you are a specialist in this field.'

It was unclear whether such comments were a joke or a barb, but faced with intermittent fireworks, John became an accomplished defuser of jumping Jacks: 'I too am delighted to think I am now a specialist in this field. It shows how easy it is to become a charlatan.'

But the tensions grew and, in 1974, I inadvertently added to them. I had often asked Jack about the early days, but he was usually too busy to enlighten me, so again I turned to John, who loved to tell (and embroider) the tales. To ensure I got the stories straight before I retold them to a new generation of students, I took four sides of notes. Jack asked to see them, and his only comment to me was, 'Very interesting.'

However, on his return to Norwich weeks later he wrote to John:

'I saw a paper [the notes] which Trevor had written as a result of a discussion with you and . . . I think he has some intention that I have not been told about . . . If this is so, I think it is ill-advised . . . I am planning to expend in retire- ment the major portion of my remaining energy on Lough Ine . . . including a book to which there would be an historical introduction. . . . I don't want the gun to be jumped.'

John replied that I had merely asked him a number of questions and jotted down the answers: 'I assumed that the information was solely for his personal interest, and any idea that he had any other "intention" – whether benign or sinister – did not (and does not) cross my mind.'

Surprisingly, Jack debited John's account more than mine, probably because in the very next paragraph John had stated: 'With regard to the future of Lough Ine I am . . . as free to discuss it as anyone else.' Although he then went on to say that co- ordinated action best served their purpose and he would do nothing unless specifically asked, he had asserted his *right* to discuss it.

Perhaps Jack interpreted this as a qualification, maybe even a threat, when what he had expected was an apology.

Jack's own history was never published and over twenty years later, when the idea of writing the story of Lough Ine *did* occur to me, I nervously wrote to tell him. To my relief, he replied: 'I think it is a great idea to try and preserve some of the joy of working in Lough Ine . . . I shall be very willing to help you all I can.'

But in 1974 he felt quite differently, and there was worse to come . . .

For several years we had studied the Western Trough in which the commonest animal was a tube-dwelling worm. Based on John's preliminary identification, we had called it *Pygospio elegans*. But when we began to draft the results for publication, Jack astutely noted: 'I am a little concerned because this animal does not seem to occupy the niche normally characteristic of *Pygospio*. This is the kind of circumstance that would lead me to look again at the systematic determination. Would F.J.E. [John] please reassert that he is satisfied that we have *Pygospio*, or alternatively arrange to send specimens immediately for verification.'

It took a year for better preserved material to be collected and correctly identified as *Pseudopolydora*, not *Pygospio*. It would

have been a serious error if we had published the work with the main actor in the story wrongly named. Jack accused John of incompetence and of not returning specimens promptly for re-evaluation.

That same year Jean Hasset, Louis Renouf's daughter, brought her father's visitors' book to the Glannafeen lab and asked Jack and John to write something to update the entry Jack had made in 1948. Jack was too busy to comply, so John wrote a one-page account of the expeditions.

Weeks later, Jack rang John to express displeasure at his entry.

John replied:

'Your phone call was my first intimation that you were upset about Jean's book and I am sorry to learn this. You are quite free to do anything you wish in the matter; I have no objection to my informal entry being deleted and replaced by a definitive version and will gladly write to Jean, if you wish . . . My sole aim was to return the book to Jean before we left Lough Ine. I thought that in writing the short piece and suggesting we both autographed it, I was performing some small service when you were busy with other matters . . .'

Unfortunately, John added a testy response to Jack's accusation that he had usurped his prerogative by using the last blank page in the book. It was a mistake. Jack responded to what he considered John's sanctimonius aspersions on his behaviour, and accused him of being heavy-footed.

'To write something in this book . . . is a very personal matter . . . I shall write more fully and without regard to what you have put. I think however that your attempt to speak for me ... would be better withdrawn and replaced by something purely on your own behalf.'

John replied:

'If it was heavy-footed of me to assume that, after so many years together at Lough Ine, our views of events were virtually identical, or sanctimonious to say I was saddened by your abuse of me, I must accept both charges ... This is not to say I do not wish to redress the hurt and I hope my letter to Jean will go some way towards this.'

But John's letter to Renouf's daughter did not heal the rift and Jack issued an ultimatum:

'There are two problems which must be resolved before our collaboration can return to its normal peaceful state. . . .

'I am not willing for you to act as my public relations officer ... A single letter to Trevor [concerning my notes for a supposed "history"] brought a full reassurance by return of post which I accepted. A similar assurance (without reservation) from you would have gone far to heal the damage you did. You must make up your mind either to accept fully the conditions made in my letter or not to come. I indeed hope you will do the former as it would be a sad ending to our long-standing collaboration if you were to refuse. . . .

'Last year there was a complete shut-down of collaboration on your part, with serious damage to the research. I am not willing to let this happen again . . .'

It was true that during the previous summer at Lough Ine John had spent an inordinate amount of time with his new camera trying to get photographs of the local fauna rather than mucking in with the projects. He tried to make amends:

> 'It did not occur to me that you had doubts about my attitude or that you were seeking a specific pledge from me not to do what I had no intent to do anyway. If, as is now clear, you need reassurance on this score then of course I give it to you unreservedly . . . I must tell you how distressed I am that after so many years of work together you continue to question my loyalty . . . and repeat my reassurance from the bottom of my heart . . .'

I knew nothing of this at the time, but I noted that in 1976 John tried to take a more active part in the scientific work and began to analyse the data from our research on the interrelations between *Codium* and sea urchins. Unfortunately, Jack misinterpreted this as an attempt to 'take over' the project and the wound between the two of them refused to heal.

In 1978 Jack again complained that John did not answer his letters and when John finally indicated the dates on which he hoped to come to Lough Ine, Jack made it clear that they were 'not convenient'.

Later he wrote to John about matters arising as a consequence of what he referred to as his 'withdrawal' from the expeditions.

> 'I am not committed to providing facilities for someone who has ceased to take an effective part in the work . . . as shown by the *Pygospio* affair, which so nearly brought disaster to our most important paper . . .'

He was also still simmering over Renouf's visitors' book, and hinted at other matters in which John had not hesitated to cause him a great deal of trouble. Sadly, the letter ended with assertion that 'Times change, people change, and your feet have grown bigger'!

John was dumbfounded, and made one last attempt to put his case:

'I do not understand why you continue to cherish a grudge about Jean's book. I fully understand how deeply you felt about it . . . That was why I wrote to Jean and asked her to remove my contribution and solicit a substitute from you. What more could I have done?

'The matter of *Pygospio* has clearly not been forgiven either. I make no excuse for my initial misidentification, which was a chastening experience and dented such confidence as I have. It was stupid. But . . . I was as doubtful as you about the identification . . . the material I had was without tails and terribly fixed in methylated spirits. That was why an agreed task . . . was to examine living worms to fix and send to other people. There was no question of my allowing the *Pygospio* identification to appear in print . . .

'In the matter of Trevor's "History" I was completely innocent. There was no conspiracy . . . Trevor was completely innocent too. The whole theory was a product of your imagination, and you even made me promise not to do anything similar before I came to Lough Ine again — analogous to accusing someone unjustly of theft and then making them promise not to steal the spoons before inviting them to dinner . . .

'There is nothing, over the years, which I would not have

done for you personally or for the success of the Lough Ine project. But I have not attempted to express your views . . . On only one occasion did I have any discussions about Lough Ine to which you were not a party, and that was when Tony Dixon came . . . I did my utmost to convince him of the unique value of Lough Ine for student field work as well as research and I can assure you I was utterly loyal to your interests . . .

'I recognise your contribution to Lough Ine now far exceeds mine. But I do not accept that I have "withdrawn" or that all the material which I have helped to collect over so many years is now free for you to publish with no more than a brief acknowledgement to me . . . Throughout most of the years of our collaboration I believe you acted towards me with reason, justice and generosity and I will not forget it, even though I now believe the situation to be the opposite. It appears that there is nothing I can do to disabuse you and it does not even seem that time can heal . . .'

Friendships endure not because of an appreciation of each other's virtues, but, rather, an easy acceptance of each other's faults. Once this erodes to mere tolerance, it becomes a conscious chore.

John never again joined the parties at Lough Ine.

LOUGH HYNE MARINE NATURE RESERVE
ANACLANN DÚLRA MHUIRI LOCH AIDHNE

Conservation

After their break-up, Jack never again mentioned John Ebling's name in my hearing. When an outsider enquired after John, Jack responded enigmatically: 'Ah, yes, I had a lot of trouble from him. He refused to put *my* laboratories on *his* maps.' I was shocked and deeply saddened by the schism.

I wrote to John to find out what had happened. In his reply, he sounded bewildered. Like him, I couldn't understand how the incidents concerning Renouf's diary, *Pygospio*, and relating stories of the early days — something John had done many times before — could possibly have led to the break-up of a partnership that had lasted for forty years.

The Bohanes had heard that the two of them had fallen out, but couldn't believe it. When they asked Jack what had become of John, he stiffened, thrust his arms down his sides as if pushing away a weight, and said, 'I have no knowledge of him.'

*

In 1974 a biologist in the Irish Electricity Board had told his chairman that Lough Ine was a perfect site for fish farming, and expressed concern that the same thought might occur to an aquaculture company. This Chinese-whispered its way to Jack as: 'The Board is intending to farm in the lough.'

Later that same year Jack met a Lieutenant Commander Jocelyn who declared he was going to set up an oyster farm in the lough, and hinted that the Government Fisheries Board intended to farm there too. It was, I presume, all possible if the government gave its approval.

At Jack's request I wrote to Máirín de Valéra who was a professor at the University of Galway. She immediately contacted the chief biologist at the Electricity Board and declared that 'Lough Ine must and will be preserved'. By the time Jocelyn's application had been turned down, Jack had consulted several Irish professors on how to protect the lough and ensure the future of his laboratories:

> 'I want . . . to make a provision in my will for the continued availability of my property at Lough Ine for marine ecological research . . . The interests of my present colleagues must be safeguarded. I am therefore looking for an organisation that would be willing to take over custodianship of my property at Lough Ine on my death. If it can strengthen the interests of conservation at Lough Ine, by helping to prevent exploitation by outside (non-local) interests, this would be an additional advantage.'

He kept me informed:

'I am concerned about the future of my property at Lough Ine, and as you know, I wrote asking the advice of Dr Went and Professor Grainger. Both are in favour of the property being vested in the Irish National Trust, An Taisce . . . I do not want the facilities just to fade out . . . I think that the most important aspect is that of continued investigation strongly coupled with conservation . . . I should like to have your comments and advice.'

I thought it was an excellent idea, and pledged my full support. Jack contacted An Taisce who enthusiastically welcomed the notion of a marine reserve at the lough. Jack wrote to let me know how things stood:

'This is a confidential matter that I will negotiate myself directly with An Taisce, and no one else is authorised to represent my views or to discuss them with anyone else. I shall, of course, be glad of any advice which you are able to provide. The establishment of a marine park is a difficult proposition, but I think that we should give all possible support if An Taisce takes it up . . . Any uncoordinated move or even unwise comment could be damaging.'

At the request of An Taisce, Jack and I prepared a case for the scientific and conservation value of Lough Ine. Jack argued that it was not only a unique site containing many rare species, but was so well studied it provided a yardstick against which future community changes resulting from climate change or environmental degradation could be evaluated.

'For the proper assessment of such changes in the world a

number of reference stations will be needed. It is necessary to know not merely what plants and animals are present . . . but also how they interact in the ecological system and what fluctuations normally occur. The number of areas about which there exists an intensive body of knowledge is relatively small. The conservation and dedication of Lough Ine to this purpose would be a most valuable contribution to a problem of international importance.'

All decisions on the natural environment are based on insufficient evidence. Yet ecologists are constantly asked, 'What will happen if . . .?' Jack believed that Lough Ine might provide some of the answers.

The local committee of An Taisce agreed, and concluded that:

'Lough Ine is worthy of being designated as a nature reserve . . . This report justifies the imposition of restrictions if these are thought to be necessary to protect the lough now or in the future.

'As it is now fifty years since Professor Renouf made his first . . . investigation of Lough Ine, it is surely an appropriate time to become really serious about conserving this unique area.'

Thanks to Jack, Lough Ine had become the most intensively studied patch of seawater in the world. In June 1981 it was made a statutory marine nature reserve, the first to be designated in Europe. Even now there are only three in the entire British Isles. Recently it has been proposed as a National Heritage Area, and is a candidate to become a Special Area of Conservation under the European Habitats Directive.

The local wildlife officer policed the lough, and John Bohane became the resident 'caretaker'. A permit was now required to do any diving or research there. An additional bonus for Jack was that no powerful outboards were allowed and the maximum speed for any craft was only five knots. You might as well row.

The benefits were soon felt. In 1986 the *Kowloon Bridge*, an 89,000-ton bulk ore carrier (whose notorious sister ship, the *Derbyshire*, had mysteriously sunk with all hands only six years earlier), rammed into the Stags Rock three miles off Carrigathorna. Cork County Council slung a straw boom across the throat of the Rapids and another across Barloge Creek. These prevented twelve separate oil slicks from entering the lough.

It had been saved because, thanks to Jack, it had been declared precious.

Fall-out

To ensure that the lough could fulfil its role as an environmental monitor, Jack decided to study ever more sites and to revisit some communities again and again. Lough Ine had provided a unique opportunity. Most research projects are funded for only three years, whereas here, because all the senior members had given their services for free, and Jack had generously subsidised the laboratories from his own pocket, our projects could be sustained over decades. Thus we observed not just seasonal changes in the communities, but the longer-term natural fluctuations that are a feature of living systems. To many ecologists, such fluctuations are just an irritating 'noise' in the system they are trying to study. To Jack they were the essence of the system, the fulcrum around which the balance of nature was poised.

We can only determine that things have changed or deteriorated if we knew what they were like before – not just in one, perhaps atypical, year, but over the range of variation that represents the 'natural' state. So we continued to monitor some communities year after year.

In 1978 Jack generously offered my family (I now had a daughter and a son) the hospitality of the labs: 'You might wish to bring Win and the children – on a special visit not coinciding with the students.'

Viv Pratt and her husband received a similar invitation and Jack was wonderful with their daughters, spending time with them and exploring the tide pools, as he had done decades before with his father. But as soon as a student arrived he reverted to his 'busy' mode.

Unfortunately, I never took up his offer, for I was invited to teach a summer school at the Friday Harbor Laboratories in the United States, and was awarded a Fulbright-Hayes Fellowship to spend a year at the University of California in Santa Barbara. This meant that I missed two summers at the lough.

Jack consulted me about his future research programme which he stressed would make full use of my expertise. He intended to carry out a general study of shallow-water rocky slopes and then the shallow subtidal mud flats, to provide as full an account as possible of the communities in Lough Ine, Barloge Creek and the open sea.

He added: 'I am not willing to accept any off-loading of work on to me, as has happened in the past. Co-authorship implies taking a substantial share of the immense amount of work which is inevitably involved. The very pleasant field work is a very small proportion of what has to be done . . .'

We had recently completed a study of the kelp forests off Carrigathorna and another of the tide pools on the shore. We were now to survey the weedy shallows in the lough on a grand scale. Jack wanted to describe every corner of the lough and make it the definitive site, but this emphasis on descriptive work concerned me. Having showed other marine biologists how to experiment in the sea, he was proposing that we devote all our energy to surveys that would not have been out of place in Renouf's time. I wrote to try to persuade Jack that we could at the same time conduct short-term experiments on predator-prey relationships, but he took my carelessly phrased suggestions as a criticism:

> 'You described what I propose as "descriptive" and of "parochial" interest. It would be irresponsible of me to support your application to the Royal Society for funds to join a programme you regard in this way.
>
> 'I have in recent years suffered from a lot of off-loading – in F.J.E.'s case, in my opinion, without excuse – in your case justified but real nevertheless . . .'

He praised my contribution to our joint publications:

> 'You did an important and essential job which I think will stand the test of time. . . . I believe these papers to be substantial and to be worth many "quick-offs" . . .
>
> 'With your perfectly understandable involvement in other activities your contribution to the Lough Ine work has greatly diminished . . . I do not wish to hurt your feelings, but I do not provide a research-by-proxy service; I am not willing to be exploited [or] to put on a quick-off programme . . . I think it is better if you missed another year. By that time we may have the opportunity of discussing the situation at leisure.

'In view of all this I shall again wait for your confirmation of willingness to act as a systematic expert before sending you a few specimens to identify.'

Dear Jack,

I'm sorry that my comments in my last letter upset you, but I felt they had to be said. However, you seem to have misunderstood some of the differences between us. I am no more in favour of quick, superficial studies than you are.

In my view Lough Ine's most significant contributions to marine ecology are the experimental papers. I think it is a pity that we pioneered what is now one of the most active fields of research and no longer participate in it. It is a great shame, both for our research and for the students, that we now do almost no experimental work.

That is, I promise, the last I shall say on the subject. I just wanted you to know how I feel. I hope that in future you will feel able to extend an invitation to me again. I would be very sorry to sever my ties with you and Lough Ine.

Meanwhile, I will of course do what I can with the identifications. I will also do whatever is asked of me to help with the projects in which I am already deeply involved.

Yours, with best wishes,

Trevor

Jack did indeed send me specimens to identify, but he never renewed the invitation to join him at Lough Ine.

Two years later I was offered the Regius Chair of Marine Biology at the University of Bergen. The international panel that assessed my research work praised the ingenuity and elegance of the experimentation — skills I had honed by watching Jack Kitching.

Alone

Jack thought it would all go on as before but, without John or me to jolly things along, the trips to Lough Ine now seemed all work and no play to many of the students.

The great paradox of Lough Ine was that, although Jack's pioneering work had propelled marine ecology into the future, he wanted the place to remain cocooned in the past, safe from the glare of electric light and the growl of outboard motors. He strove to preserve what he loved about it – its other-worldliness. And who could blame him?

Jack was still getting his clothes from the dressing-up box, still defying medical thinking by subsisting on lashings of lard and sugar, and crisping in the sun each summer. He was the same as ever, but the students had changed. They had less enthusiasm to share the claustrophobic world he had so carefully preserved. His expeditions were stuck in the 1950s, when joints were for carving not smoking, when students were 'good chaps' and 'grand girls', content to Swallows-and-Amazons away their summers, mucking in with the tasks and singing 'Polly wolly doodle all the day' for Sunday entertainment. Now long retired, he had little contact with undergraduates back at the university, and his 'Call me Jack' could never bridge the gulf. He had lost the company of those with whom he always felt most at ease.

Always having been spared the domestic running of the trips, he now resorted to ten pages of instructions for the students, plus rotas and long lists: 'Tents are pitched on the south side of the building — leave the north side free for J.A.K.'s tent, and *the rest of the north side empty.*'

He had always camped separately from the rest of the party, but now that he was alone this emphasised his isolation.

'The Elsan chemical toilets must be emptied twice daily — midday and shortly before supper — these times should be adhered to. This duty will be undertaken by the men of the party in turn in accordance with a list and with the authorised procedure . . .'

I was told that the men piddled in the bushes so as not to fill up the toilet cans, and resented Jack for not doing so.

Jack was never good at briefing people for the task ahead, and when they got things wrong he got irritated and so did they. He felt that he was surrounded by incompetents and subversives and accused one student of mislaying thirty-six metres of rope. She was certain she had put it in the corner of the loft.

'Then where has it gone to?' he asked.

Later John Bohane found the rope in a sack, exactly where she had said.

Rats got into the hut, and Jack decided to disinfect the entire place. Six gallons of some strong brew were ordered from Fullers, but they vanished.

'They were left on the pavement outside Donelan's pub,' Jack insisted indignantly. 'And they have been stolen!'

'I'll be hoping nobody takes a swig by mistake,' said John Bohane. 'It's fierce powerful stuff.'

He was despatched to buy another six gallons. Later the original cans turned up in the lab, and Jack had enough to disinfect the whole of County Cork.

He tried to keep all the rituals alive, and also invented new ones. Tuesday became the day for ordering meat from the butcher in Skibbereen. It didn't matter that someone had to go into town on the Monday, meat could not be ordered until Tuesday, so another trip would have to be made.

There was now more formal teaching at Lough Ine than before, and some evenings Jack would launch into lectures lasting almost two hours. The students invariably nodded off in mid-talk, and sometimes so did he.

The university department back in Norwich was concerned that their students were in an isolated spot without a telephone and with a septuagenarian in charge, so they paid a young research student to go along to save Jack from the heavy work. I don't know whether Jack considered this a sensible precaution or just unnecessary interference, but when the students clambered over the rocks to Carrigathorna, he insisted on shinning down the cliff first so that he could stand at the bottom to catch them should they fall.

His gentlemanly acts were resented. He declined to teach the women to row, thinking they should be rowed around by the men. But the biggest source of conflict was evening trips to Skibbereen. The students no longer tolerated the seclusion of the campsite and Jack had never understood why anyone needed to go to a bar. After one booze-up a student developed such a hangover he refused to get up next morning. Thereafter, Jack banned all visits to the pub – but they went just the same and put up with the tense atmosphere it created. Worse still, some bought bottles of gin and would swill down tumblersful on empty stomachs before dinner, then pass out on the grass outside Jack's tent.

The more Jack laid down the law, the more he became isolated. At meal times he would sit at the head of the table and the seats would fill up from the other end. I have often wondered whether, if I had been there, it might have made a difference.

The students' *unofficial* diary records: 'The revolution had begun. A small party of troops . . . smuggled supplies in to the freedom fighters . . . a bottle of gin, a bottle of vodka, a bottle of Dubonnet, a bottle of orangeade (for Mandy) . . . M set up a diversion by taking Jack sampling.'

They would not have believed that once it was the best party to which I was ever invited, and that everyone from the early days remembered only good times, when students 'awoke depressed at the thought of leaving Lough Ine'. A typical recollection from the 1950s was that 'packing up camp was always sad. Standing on the hill in the early morning with a flat calm on the lough, one always vowed to come back. I shall be forever grateful to those good friends who introduced me to Lough Ine and whose company I shared there.' Others recalled 'all the wonderful camaraderie that lingers still in the memory after forty-five years', and described Jack as 'a really great man to be with, inspiring us with his enthusiasm' and the 'remarkable, vital person I feel privileged to have spent days with at Lough Ine'.

Although An Taisce had risen to the challenge of making the lough a nature reserve, they were not interested in taking over the laboratories. So in 1979 Jack approached his old department in Bristol and invited Dr Colin Little to carry out research at the lough. He hoped that this would infuse Colin with enthusiasm for the place and thus persuade his university to adopt the laboratories.

On his first trip, finding no sign of life at the lough, Colin drove towards Baltimore and met Jack with a file of students trailing behind. They had been trudging across Cape Clear, and were now wearily walking home.

'Would you like a lift?' Colin enquired.

Jack was appalled. 'No, thank you. I need some exercise,' he said brusquely and vanished up the road at top speed.

Several exhausted students piled into the van.

Being always so fit himself, Jack interpreted any reduced effort as malingering. Even when Professor Dixon, who had the future of the expeditions in his hands, had been laid low with a slipped disc, Jack was unsympathetic.

He had retained his ability to astound by ladling syrup on to his bacon. He once covered a slice of currant bread with peanut butter, golden syrup, raspberry jam and Bengal hot chutney, describing the flavour as 'interesting'. He still had a flair for naming things; rough scones made with coarse Irish flour became 'knobbly crustose'. But the once-surprising 'Jack in a box' was now trapped in a box from which he could not escape.

There were still lighter moments, as when the diving party made Christmas pudding with a strange-tasting custard. A label found submerged in the custard revealed it had been made with the water used to boil the pudding tin. It was a long way from John Ebling's salmon meunière followed by coffee and chocolate soufflé.

At one point, there was a long dry spell, and a party was

ordered to drive to Skibbereen to fill jerry cans from a stand pipe. As they were about to depart it began to hammer with rain, but Jack would not rescind the order, so they stood for over an hour collecting water during a torrential thunderstorm, and risked injury sliding down the muddy path to the Goleen hauling huge cans full of water.

Colin Little returned the next year and John Turner and Julia Shand organised diving teams from Bristol to finish the work on Whirlpool Cliff. Jack sat patiently in a boat above and received the samples they brought up. He amused the divers by ingeniously getting them to release handfuls of sea lettuce underwater and using its sideways drift to estimate the current speed.

But there were dozens of minor conflicts. One diver emerged into the shallows and trampled through a pristine area where Jack was monitoring the growth of kelp. Jack yelled, 'Go back!' but the diver came on, dragging the precious kelp with him. He was never forgiven.

Jack was tired by suppertime and often went to bed as soon as washing up had finished, and the rest of the party drove to the pub. So there were no after-dinner chats, and no one said, 'What about tomorrow's programme?'

In 1985 even nature added to the misery. 'Sheep ticks are everywhere in the bracken,' Jack wrote. 'More numerous and worse than I have ever known them.'

Perhaps to his relief he ran the last student field trip in 1986, forty-eight years after his first.

Although Colin Little continued to come, to work on the movements of limpets, Bristol University did not have sufficient field biologists to warrant taking over Jack's laboratories. Having always sworn he would never leave them to the University at

Cork, he now asked the Professor of Zoology there if it would be interested.

In October 1987 Cork officially took receipt of Jack's land and laboratories and opened a new two-storey building on the site of the old Renouf labs on the south shore. To supply it with power they had been permitted to sling a cable over the Rapids. There was a handing-over ceremony on the grass outside the new laboratories, and Jack gave a speech.

Charles Haughey, the Prime Minister, had planned to attend the ceremony, so there had been a police presence in the area for several days. The day before the opening, two of Colin's students went snorkelling in the shallows and were accosted by the Garda. It took some time to convince them they were not terrorists.

They spent the next night outdoors lying on their stomachs during a downpour and a howling gale, monitoring the movements of limpets with the help of dim, red-light torches, so as not to disturb them.

The same policeman discovered them. 'Well then, you're not going to tell me you're marine biologists this time, are you now?'

After they explained what they were up to, the policeman crossed himself and said, 'Be Jasus, nobody could make up a story like that.' And with a despairing shake of his head, he returned to the patrol car.

Jack left without revisiting the Glannafeen laboratories. Colin Little saw him go: 'He was rowed away in state. He looked as deadpan as usual, but I wonder if he knew he would never come back.'

Jack never did complete his planned survey of the shallow-water communities, although the work kept his assistants supplied with animals for taxonomic studies. Viv Pratt worked on the small creatures for him and was excited to find what she thought was an unknown species of worm. When Jack sent this for confirmation by an expert, he made no mention of Viv's involvement. At one stage it even looked as if the new species would be named after Jack. This was inexplicable, as he had always been meticulous over such matters and generous towards junior colleagues.

The publications became fewer and thinner and appeared in minor journals. Jack sent a batch to me and on the top of one he had written: 'Just a "Quick-off" – Best wishes, Jack.'

Endings

Jack and John met again only once, at a symposium in Cork in 1990, on the research at Lough Ine. They and I were invited speakers. Although it was a relief to John to have an invitation and an excuse to visit the lough, he was nervous about meeting Jack again. He needn't have been, for Jack was no longer formidable. The contrast between them could not have been more marked. John was as ebullient as ever, but a stroke had stolen Jack's vigour. He looked and sounded frail, a ghost of his old self. I feared his lecture might be a disaster, but on stage he rose to the occasion and spoke well. I saw John helping him across the road. They were chatting, perhaps about old times.

Bunny, my old supervisor, was missing from the reunion. She had died in 1986 after a long struggle with cancer. I felt suddenly older when I wrote her obituary:

'She was a warm but private person . . . generous with her

time and her resources. I remember that fearsome disapproving look when, as usual, I had done something wrong, or her face convulsed, laughing until she cried, at something I had said. But most vividly I recall her dressed in a flimsy raincoat, lashed by wind and rain, attempting to survey a shore, heroically oblivious to the weather.'

John had retired from the University of Sheffield in 1982, but his expertise was too valuable to waste and he was given a research laboratory in the Royal Hallamshire Hospital. The *Textbook of Dermatology* he co-edited ran into its fifth edition.

The Eblings decided to return to Lough Ine again in 1992, and Erika booked the trip. But it was not to be. The next day they went into town together to buy photo albums for their anticipated holiday snaps, then John went off to work. After a lively lunch with friends he returned to the hospital. Suddenly he raised his hand to his chest and keeled over in the middle of the concourse. His research assistant walked past and saw a knot of people gathering, but never guessed they were clustering around John. Within reach of a cardiac unit John's heart had failed him. He was seventy-four.

On the day of John's funeral my daughter was sitting her finals at the University of Wales, in Bangor. She was required to write a summary of a scientific paper on which the authors' own summary had been obscured. The paper was written by Kitching and Ebling.

At the memorial service to John the eulogy recalled some of his favourite openings:

'Consider for a moment the entire history of the world . . .'

The obituaries caught the essence of him:

'Few have the ability to assess the significance of the particular in the context of the whole. John Ebling was a remarkable exception and his career encompassed almost all aspects of biology, from marine biology to clinical dermatological practice. He was equally at home in discussions on social hierarchies, global warming or hirsutism.'

'John was an enthusiastic and inspiring teacher . . . whose insight guided his students along creative avenues of research. Without detracting from the serious and often profound aspects of research, he was able with his particular sense of humour to draw amusement from sometimes the most unlikely of sources.'

John Bohane put it more simply: 'He was a grand man who loved a joke.'

Erika was pleased to receive a letter of condolence from Jack.

Jack was awarded a well-deserved honorary degree from the National University of Ireland in recognition of his work at Lough Ine. He had begun to write a book on marine ecology, but the effort became too much for him and he asked Colin Little to take it over. Another stroke temporarily paralysed his leg. I was told that his voice had faded to a whisper, and so had his writing. It had always been small but now it was minute, the tracks left by a tiny lost spider. When I wrote to tell him I was writing this book and asked if he had any papers that might help me, the spider replied: 'I'm sorry I can't be more helpful, but I am old and both my memory and my handwriting get smaller. Any notes I have are lost because there is nowhere safe from being swamped.'

I was saddened by the tone; it was difficult to imagine Jack so enfeebled. In 1993 both he and his wife entered a nursing home. Evelyn was still fairly robust, but was suffering from Alzheimer's disease. She thought the watercolours on the wall were family snaps – 'Since that was taken we've finished the building work.'

By the end of 1994 Jack, who once bestrode the hills, could no longer walk nor stand. He who had kept peanut butter factories in full production could not feed himself, and the simplest meal took an age to consume. He did not take kindly to being dependent on others, but seemed resigned and sadly content with a task completed.

Then, when his wife fell and broke her hip, he sank into silence and watched the trees discard their leaves. Only one bright moment remained. He received a copy of the textbook he had started and Colin Little had finished. He spent many hours looking at it, slowly turning the pages as if they were the pages of his life.

It was John's death and Jack's enfeeblement that caused me to write this book, for now there was no one else who knew the story. Before it was finished Jack died in his bed at the age of eighty-seven, exactly sixty years after he had first seen Lough Ine.

Anniversary

For our silver wedding anniversary in 1993, Win and I toured Ireland, now a very different Ireland that was going places in the modern world. Almost every village wore its 'Accident Black Spot' sign with pride, and the roads were as patchwork as the fields . . . 'Slow Uneven Surface' – as if Ireland had ever had any other.

Even remote districts were designated as a 'Community Alert Area' to deter the 'gougers and villains' who marauded out from the city to snatch a few valuables. 'Now everyone draws their curtains and locks their doors,' our landlady told us sadly, 'but if they want to get in they'll dance over any ditch.'

Weapons were now used more frequently, but not everywhere: CAHIRCIVEEN GUN CLUB – SHOOTING STRICTLY PROHIBITED. Drugs were a problem too. A farmer renowned for his piety was found to be feeding 'angel dust' to his cows. Only he knew why.

Gone were the Irish who once adorned the *National Geographic* with weathered faces thinking of something useful to do with seaweed. Instead, entrepreneurs squinted into the future with an eye to the main chance. Still, only one farm in ten was viable – although, we were told, generous subsidies were given to those who pledged to use less fertilizer, clip their hedges and eschew leaded paint.

Ireland was prosperous at last. It had the fastest-growing economy in Europe and was a thriving financial centre, although not everywhere: BANK OF IRELAND – OPEN TUESDAYS ONLY, 10.30-12.30. A poster in the window assured customers that even 'Jesus Saves'.

Every lane was clogged with tractors, ever on the roads, never in the fields, agricultural actors always between engage-

ments. But the lanes still flickered with wagtails, and the verges still nurtured montbretia and bees tipsy with nectar.

Echoes of old Ireland remained in the petrol pump that claimed to deliver Murphy's stout, and the surprised eyes of oncoming drivers who seem never before to have encountered a car coming round a corner on the correct side of the road. This was only one of the perils on the highway: DANGEROUS COWS CROSSING.

The girls incarcerated in the Nana Nagle Presentation Convent hung from the windows shouting, 'Love your T-shirt, laddo!' to the passing talent. But many convents had closed as novitiates deserted to the disco.

The monasteries were also emptying. 'I find the life of solitude is not for me,' a tyro monk confessed.

'Oh, really,' said the Bishop sadly. 'What's her name?'

In Ballinspittle a statue of the Virgin Mary had been seen perambulating around the town and the local police sergeant reported that she took off and hovered in mid-air. The local authority erected a sign warning that it took no responsibility for road collisions in the vicinity of the statue. But in Ireland, apparently, spectres haunt many a road: GHOST ISLAND AHEAD.

At Cahir we asked an old lady where she would recommend for a good cup of coffee. 'Oh,' she said. 'I suppose Clonmel's too far for you?' It was ten miles away.

We ended up at Skibbereen, where the narrow streets were crammed with traffic. The carts and cowpats that were conspicuous fifteen years before had been consigned to history. The donkey was now officially designated an endangered species.

Lots of emigrants had returned for 'Welcome Home Week' to enjoy the Cork Ideal Housewife competition, and see who would be crowned 'Maid of the Isles'.

West Cork was becoming a trendy place for 'blow-ins' from abroad. Fields, the grocers, was now twice as big, and where once it had stocked just staples, now there were truffles, venison and ten kinds of olive oil.

But Donelan's pub looked exactly the same. On the wall were curled and fading snapshots of Jack, John and Erika, looking as young and fresh as the students around them.

Ernie Donelan had died, and his son Kevin now ran the coach and taxi business. Ernie's widow, Mary, still manned the bar in her social welfare centre that served alcohol. In the corner was a lady in her nineties. She looked twenty years younger but her mind had waned, so Mary took her in twice a week to give the son a break. The old lady seemed quietly content, but frequently asked, 'Is David here yet?'

'He'll be back soon,' Mary replied, and that seemed to reassure her for another quarter of an hour.

Mrs Donelan gave away more drinks than she sold and placated a swarm of grandchildren with Coke and crisps. 'It's not like it was,' she complained. 'The *Garda* are terrible strict now on closing times.'

She had taken a lodger – a weightlifter – in the flat above the bar, but one night he and his weights came through the ceiling. The pub had to close for repairs and Mary was shipwrecked without a job or a place to be. The mended ceiling was repainted to match the nicotined walls. 'We had a divil of a job getting the colour,' she confessed.

She bought us silvered coasters for our anniversary.

The road to Lough Ine was just as I remembered it, except for the Limbo Riding School. I recalled those long hikes from Skibbereen so many summers ago. The hedgerows still bustled with bee-busy foxgloves and the stately plumes of royal fern, and the lough still sparkled in the intermittent sunshine. Along the road beside the Western Trough the wind interrogated the trees. Leaves swirled and hid in the shadows, whispering among themselves. Incredibly, it too was a Community Alert Area:

LOUGH HYNE MARINE NATURE RESERVE

Beneath the sign was the little gate that led to the quay in the Goleen. The postbox where once we left Guinness for Willy Nilly the postman was now a rotting carcass inhabited by bracket fungi.

Long ago Sonny Donovan had planted trees at random here. 'They'll never grow,' John Ebling had said, but now, tall and tangled, they made a dark tunnel of the steep path.

It had finished sadly for Sonny and his strange, wild sister, Kathleen. They both ended their lonely lives in care, and she died after falling down the stairs in an asylum.

The quay below looked as good as new, and we sat on it for an hour or more. I searched the shallows but the purple urchins were gone; a man had been arrested for filching them to sell at £3 a time. Otherwise it was a place untouched by time. Win sketched the rocks – some still bore the traces of the yellow crosses we had painted years before.

The next day we visited the Bohanes. Time had been kind to them, and John was still full of tales. The police had caught smugglers who had hidden drugs in the hedgerows. At about this time John had bought a brand new car and Philomena a new outfit. He then fell and hurt his leg, which everyone began to pull.

'Ah, it's terrible about the leg, John, but be thankful the *Garda* didn't catch you.'

Neilly found a message in a bottle washed up on the Coosh. It had been at sea for four years after being tossed overboard from an American liner. He replied to the shipping line and the post brought four tickets for a Caribbean cruise. Neilly and his wife Annie went, and their family drew lots for the other two places. They rode in a limousine through Miami in the St Patrick's Day parade and had the time of their lives. 'It was grand,' Neilly confessed. It was the first time they had been abroad.

'It's all changed,' they told me sadly. 'There's no gossiping now.' Bigger milk churns meant fewer trips to the creamery, and the new separator never broke down. 'It's there and back with hardly a hello from a neighbour. And the telephone company's cut the length of local calls so there's no daily chatter with friends any more.'

The Bohanes still thought kindly of the eccentric expeditions that once came to their lough. John was working on Jack's brown boat. He had put in a new keel and replaced some planks. It was sixty-four years old and would soon be back in the water. He walked down with us to the Dromadoon lab overlooking the Rapids. Inside it smelled of damp and musty paper. The stools were riddled with woodworm. John shook his head. 'Anyone with a wooden leg would be in trouble sittin' there, I'm thinkin'.'

I wondered how many biologists would ever sit here again, whatever the nature of their legs.

Outside, among the bracken crosiers, we found a dead sheep, once a fine animal but now just withered skin and a snake of vertebrae, a mummy callously unwrapped for public display.

We borrowed John's boat and rowed through the mist across to Glannafeen. The Rapids Quay was beginning to crumble, the paths had healed without trace, and the field where once we had camped was now halved in size by the encroaching bracken. On the grass in front of the laboratory was one of the black boats, its bottom stoved in. Two years earlier someone had left the fibre-glass boat poorly tethered, and it had been sucked down the Rapids, never to be seen again.

The laboratory roof was felted with yellow lichen, a window pane was broken and rats had gnawed away the foot of the door. The exterior, once repainted gleaming white each year, had peeled to expose the bruised red walls beneath. A sign over the door said: THE KITCHING LABORATORY. Thankfully, Jack was not here to see the drab shadow of his dream.

Inside the mess room the soft touch of time had left memories in every corner. John Ebling's boots were still in the washroom, his recipe cards on the shelf. My Sunday songs nestled unembarrassed in the filing cabinet. Jack's old macintosh hung behind the door, a crumpled record card still in the pocket. A small brown box held the cards for those last unfinished jobs – he had almost made it to the four-thousandth task.

Holding one of those oh-so-familar task cards in my hand was like finding a key to the past. For a moment it was as if I was there again telling ghost stories by candlelight to a rapt audience.

Students from Cork had little use for Kitching's lab, but others came from the universities of Bristol, Newcastle and

Wales, still diving, still experimenting and, I hope, still falling in love with the place. Jack and John would be pleased they came. But our Lough Ine had gone for ever.

I savour the present, and rarely revisit the 'good old days' of which we pretend the past was exclusively composed. But there in the deserted hut I longed for those lost summers of long ago that were inhabited by so many bright and cheerful students – and two of the most extraordinary people I have ever known.

Win and I rowed around the lough on a soft, mother-of-pearl morning. A heron stood on a rock watching our approach. Then it opened its immense lazy wings, swooped low over the water, and was lost in the mist.

Ghosts

We returned to the lough that night and sat on the north quay. A cool darkness had settled on the trees, beneath us lay beds of sinister seaweed, and a reluctant moon emerged from the clouds to burnish the surface of the water.

My thoughts turned to Jack and John, both atheists with nowhere to go, nowhere for John to repeat his stories or Jack to disbelieve them.

The laboratories on the far shore were lost in shadow. But then there was a tiny flicker of light – a torch, or perhaps a candle. I could hear voices far away and faint, yet as if someone was whispering in my ear. Jack's hoarse commands, an explosion of laughter that could only be John, murmurings of the wind in the trees, and a sound I couldn't decipher. Could it be snoring? Had Jack fallen asleep?

Then the voices fell silent and the point of light died.

*

All that effort over all those years and what remains? A few neglected buildings and the ghosts of lost paths? No, far more than that. The scientific papers from Lough Ine will be read for as long as biologists tread the margins of the sea. Paul Dayton, a famous American ecologist, wrote to tell me he had visited Ireland and made a 'pilgrimage to Kitching's little lab. For an old atheist like me it was about as close to religion as I could get.'

There are no simple commandments that govern marine biology. Living communities are complex, and glib answers will not solve environmental problems. Jack and John gave us the key to dismantle the complexities, and Lough Ine was the perfect living laboratory from which to unlock the secrets of the sea. Such single-minded investigations of one site over so many years will never happen again. Both the researchers and their expeditions to Lough Ine were truly unique.

No doubt we are more efficient and effective scientists nowadays, but the pressure to secure grants and publish papers in time for the next audit is a form of instant gratification rather than philosophical contemplation. In the rush to catch the bus we may omit to check that it is going in the right direction. Jack took his time – too much time on occasion, but he wanted each study to be complete, to tell a satisfying story. And, as a result, he told good stories. Excellent research at the lough continues, but there is no place in modern ecology for the leisurely and pleasant world he built, and science is the poorer for it.

Almost every student who was privileged to explore these shores with Jack and John has been touched by the place. Thirty of them found their mates here. Several, like me, are now Professors themselves and practise the skills they first learned here under such expert tutelage. We are still investigating those parts of life that are knowable, the parts that do not involve the behaviour of human beings.

And the lough remains . . . in all its moods, as magical and mysterious as ever, still awaiting discovery, for it holds far more secrets than we uncovered. And, as the ancient Celts knew . . . the sea is the beginning and end of all things.